Coproduction in the Recording Studio: Perspectives from the Vocal Booth

Coproduction in the Recording Studio: Perspectives from the Vocal Booth details how recording studio environments affect performance in the vocal booth.

Drawing on interviews with professional session singers, this book considers sociocultural and sociotechnical theory, the modern home studio space, as well as isolation and self-recording in light of the COVID-19 pandemic.

This is cutting-edge reading for advanced undergraduates, scholars and professionals working in the disciplines of recording studio production, vocal performance, audio engineering and music technology.

Rod Davies is Lecturer in Popular Music at Monash University, Australia. He is an experienced session singer with a research focus on creative collaboration in music.

Coproduction in the Recording Studio
Perspectives from the Vocal Booth

Rod Davies

Routledge
Taylor & Francis Group

LONDON AND NEW YORK

First published 2022
by Routledge
2 Park Square, Milton Park, Abingdon, Oxon OX14 4RN

and by Routledge
605 Third Avenue, New York, NY 10158

Routledge is an imprint of the Taylor & Francis Group, an informa business

© 2022 Rod Davies

British Library Cataloguing-in-Publication Data
A catalogue record for this book is available from the British Library

Library of Congress Cataloging-in-Publication Data
A catalog record for this book has been requested

ISBN: 978-0-367-70548-0 (hbk)
ISBN: 978-0-367-70551-0 (pbk)
ISBN: 978-1-003-14686-5 (ebk)

DOI: 10.4324/9781003146865

Typeset in Times New Roman
by Apex CoVantage, LLC

Contents

Illustrations

Figures

Tables

Participants

(AD)	Andrew De Silva
(AL)	Angela Librandi
(CS)	Carmen Smith
(DP)	Darren Percival
(GC)	Glenn Cunningham
(GP)	Gary Pinto
(HT)	Hayley Teal
(JC)	Jacinta Caruana
(JD)	Jaydean Miranda
(JF)	Jud Field
(JM)	Janine Maunder
(JMc)	Jade MacRae
(LF)	Lindsay Field
(LM)	Lisa Maxwell-Cripps
(MS)	Michelle Serret-Cursio
(NF)	Nina Ferro
(NN)	Nikki Nichols
(SA)	Susie Ahern
(SG)	Susie Goble
(SW)	Steve Wade

Acknowledgements

I want to thank all of the participants in this study who coproduced this book with me. Thank you for sharing openly and freely your creativity, intelligence and experience to create what I believe is a unique text. To these hidden musicians, I offer my respect and gratitude for everything you do to sustain a wonderful working environment in one of the most enjoyable jobs in the world. I also thank the producers, engineers, musicians and songwriters, without whom it would just be one big acapella singalong.

1 Introduction

Aims and focus

This is an untold story. A story about recording studio production, which in recent years has grown as a field of research but has not yet borne clear witness to the voices of the singers behind the glass. This book aims to help us understand what session singers do, within the scope and context of the Australian music industry between the late 1970s and today. I examine how changes to technology and recording environments have changed the nature of session singing, and how social and cultural changes have marked even greater shifts in recent years. By foregrounding the traditional organisational system of the recording studio with contemporary testimonies of session singer experiences, we may see clearly the way new technologies and new expectations have shifted the old framework and created a new type of coproduction, challenging former expectations and asking us to reconsider how a work in the recording studio is both coproduced and acknowledged in a contemporary cultural, social, technical and financial sense.

This book does not interview producers or engineers. It does not claim to represent perspectives from *both sides of the glass*, but rather a clear and unobstructed vocal booth perspective. It is my hope that when you are finished reading, you will be informed and challenged by some of the ideas presented. You might agree or disagree, but above all, I hope you will consider these perspectives when you form your own ideas about what coproduction in the recording studio looks like today.

The thesis of this book is as follows:

* Expectations of roles, responsibilities and processes within the recording studio system have been built upon traditional notions of coproduction. Yet even within the traditional system, ambiguities exist in relation to roles, responsibilities and processes.
* Change that was driven by technology in the late 1990s was the first major shift towards combining these roles and responsibilities and thus changing the processes of coproduction.

DOI: 10.4324/9781003146865-1

- The COVID-19 pandemic of 2020 has been the second major shift, challenging an array of expectations, roles, responsibilities and processes. However, many of these changes have not been acknowledged. Up until now, they have remained hidden, invisible to people outside the domain, and muted in tacit understanding to those inside the domain. This was not a technological change but rather a societal transformation that precipitated a cultural change.

To most outsiders and many insiders, perspectives from the vocal booth are ambiguous, unwritten and unexamined. This book aims to change that, not by prescribing what coproduction should look like or what the outcome of negotiations in the recording studio should be, but rather by equipping participants with knowledge that can assist in making the practices fair, effective and efficient. In so doing, it will hopefully interrogate the question 'what do session singers actually do?'

By exploring this question in depth, and thus investigating the various types of coproduction that emerge from the discussion, conclusions may be drawn regarding what the implications might be for models of industry and education.

Scope

Defining recording studio production

In order to understand the recording studio domain, one must understand its cultural fabric, which is based on a particular organisational system and long-standing social norms. Cultural practices in the recording studio have been largely based on traditional roles that individuals have played in the recording studio environment over many years, including a producer, an engineer, musicians and songwriters or composers, who work together to coproduce a musical outcome.

At the centre of the collaboration is the producer, a role Howlett (2012) describes as a 'nexus' through which the work of the artist, technology and commercial interest come together. Moorefield (2005) describes how this perception has grown over the years and was catalysed by the emergence of individuals such as George Martin, Phil Spectre and Brian Eno, whose work greatly influenced modern popular music and culture. Most of the credit for creative decision-making in the recording studio is given to the producer, but it would be false to suggest that other participants do not contribute to the dissemination of ideas as well.

The studio engineer controls the tangible technical aspects of a recording session, a role that became more and more specialised as technology advanced throughout the 1950s and 1960s (Horning, 2004; Théberge, 2004). The engineer works with the producer to 'deliver particular styles of sound on a recording' (Watson, 2014) and to do it efficiently and *instinctively*. The role might involve placing microphones in optimal positions in the studio along with appropriate sound dampening, routing audio signals between recording spaces and control rooms (both tracking and headphone monitoring), managing effects and software plug-ins, and achieving consistent sound levels on input (recording) in order to preserve sound quality and fidelity, and output (monitoring) in order to optimise and preserve participants' listening and hearing. In addition to the tangible technical aspects, engineers should also have good listening skills and be able to communicate clearly with the producer in terms of *aesthetics* and to effectively 'engineer the performance' (Horning, 2004).

The musician makes up the traditional *triumvirate* of recording studio personnel by providing the musical sounds to be recorded. In the traditional notion of this organisational system, musicians are present with the producer and engineer in the recording studio, taking direction from the producer and performing *takes*, which are captured by the technologies operated by the studio engineer.

In addition, there are other stakeholders who have direct interest in the work, namely publishers, record companies, advertising agencies, songwriters, recording artists, and other clients who may range from private individuals to large corporate enterprises.

While it is useful to assign labels to these individuals (as they appear on the credits of an album sleeve), in practice the delineation of these roles is less than clear (Davis & Parker, 2013; Moorefield, 2005). Mellor (1996) identifies the roles of 'engineer-producer' and 'musician-producer', while there has also been a modern emergence of the producer as composer or songwriter (Moorefield, 2005). As Driver (2015, pp. 44–45) remarks, 'analysing the collaborative mix in the recording studio by attempting to analyse discrete individual roles is problematic, as these roles are flexible and blurred in practice'.

One reason for this ambiguity is that the recording studio is often a site of contestation of ideas between its participants (McIntyre, 2008a) and how these struggles play out can be decided by the various degrees of capital each participant controls (Bourdieu, 1990). *Cultural capital* is derived through an understanding of a symbolic system of conventions, knowledges and techniques that are required to be a creative contributor in the environment (McIntyre, 2008a). Usually, the producer and engineer earn their roles as a result of their high level of cultural capital. Musicians are often valued just as highly for their *social capital*, which is determined by how 'good' they are to work with. Publishers, record companies, artists and advertising companies are often the chief stakeholders in the project and therefore typically hold the greatest *economic capital*. And finally, in each role, there is the possibility that an individual's 'celebrity' attaches an amount of *symbolic capital* to the equation.

In most scenarios, the holder of the greatest economic capital will hire an individual who holds a large amount of cultural capital to facilitate the recording process. This person is usually the producer, whose role is to lead the coproduction, bring all the pieces together and provide final oversight so that the best outcome can be achieved. To do this, they determine the other participants who are required, which includes the engineer and musicians.

Session musicians, who are often the last in the chain of 'hires', account for a large proportion of production within the music industry worldwide. As Williams (2010, p. 59) points out, 'much popular music is in fact made by unknown, unidentified musicians, hired collaborators who work out of the public eye in the recording studio or in the shadows of the concert stage'. These hidden musicians include session singers, who are a 'specialized group of singing professionals skilled to perform in the studio' (Campelo, 2015). Often their task is to record, from 'first sight', a song that is yet to be defined, which must be learned, developed and performed, all in the one recording session.

This book focuses on session singers because firstly, this rare skill is not widely understood, and secondly, because as we shall see, this is just one of the many skills a session singer brings to the coproduction process. When these hidden performers step into the recording studio they become the centre of attention, integral to the sound, the way the message is communicated and the overall aesthetic of the final outcome. As Williams (2010, pp. 63–64) states:

> They must be malleable, moving from unobtrusive scenery to the center of attention and back again. They must shadow and support, or jump-start, initiate and generate momentum and excitement. They must simultaneously project, and be devoid of personality . . . Yet, freelance musicians must be able to deliver more than the expected right notes, the most successful musicians deliver the unexpected, the execution of particular and unique musical choices that define an identity to employers and to fellow session players.

Defining the three 'waves' of recording studio production

While identity and an advanced skillset have always been integral to the session musician's *toolkit*, technological, social and cultural movements over the past 30 to 40 years have changed the way this toolkit is applied in the co-production process. In the sections below, I suggest definitions for three 'waves' that symbolise these movements.

The first wave – larger purpose-built studios

Professional studios of the 1950s and 1960s were full of new technologies that relied on skilled staff. Sound engineers were trained professionals who operated the mechanics in concert with the acoustics of large reverberant spaces that were designed 'to "capture" a "natural" performance'. (Bell, 2018, p. 15). Studios were 'formed for their acoustic properties' (Gibson, 2005, p. 193), which required large physical spaces and economic resources to do so (p. 197). These were 'highly regimented and bureaucratised institutions' (Leyshon, 2009, p. 1319), which established professional organisational structures to match the high standards of the physical and technological structures.

By the late 1970s, technology had escalated rapidly with previous technology supplanted by 'more compact and replicable successor(s)' (Bell, 2018, p. 20). The release of the *digital reverb* dramatically changed both recording practices and the architecture of the studios themselves, with new approaches to 'divide and isolate' the recorded sounds giving rise to relatively smaller and more isolated recording spaces. Yet, throughout these major renovations, notions of specialists and custodians of the recording practices remained.

The 1970s, 80s and 90s were a boom time for session musicians in Australia, which was due in large part to the laws regulating advertising. All products advertised on Australian television and radio in this era were required to be recorded in Australia using local producers, engineers and musicians. This included advertisements from the US and UK, which would be 're-voiced' before going to air in Australia. Budgets were large and work was plentiful, but in 1998, a change in Australian government policy spelled the beginning of the end for this lucrative industry. New legislation (Commonwealth of Australia, 1998) permitted the 'parallel importation of non-pirated copies of sound recordings' (Papadopoulos, 2000, p. 340), and opened the way for recordings made overseas to be broadcast on local media, a decision that decimated the local producers of content and made it much more difficult to sustain a thriving session music industry. But even bigger changes were yet to come.

The second wave – home studios

As far back as the late 1960s, people began establishing do-it-yourself (DIY) or home recording studios. This was largely considered a 'lo-fi (low-fidelity) movement, which continued to value the ease of access and ease of use over quality' (Bell, 2018, p. 18). Over the next two decades, the gap between quality and affordability diminished, and by the time the ADAT (Alesis Digital Audio Tape) arrived in 1991, the 'democratization of the recording process . . .' had begun, enabling 'the project studio market to blossom and to produce (close to) professional quality digital audio' (Burgess, 2014, p. 131). It was the 'first affordable audio device to lay claim to professional-quality sound' (Cole, 2011, p. 450), yet it was soon overtaken by the Digital Audio Workstation (DAW), which allowed home recording studios to 'have as much recording capability as the Abbey Road Studios' (Krueger, 2019, p. 194).

During this time, while larger commercial recording studios serviced the high-end of the market in Australia, home-studio producers looked to smaller local market productions, such as corporate soundtracks, radio mashups, low-budget album recordings and song demos. While perhaps not as lucrative as the decades that had preceded it, the changes brought by the second wave of the 1990s granted important opportunities to a new generation of session singers.

The third wave – remote recording

Like the second wave, the third wave has its genesis in home recording. Low-cost, high-quality digital equipment has now enabled a generation of musicians to self-sufficiently develop skills and record their own music at home. 'The DAW in its current form has signalled the dissolution of the technological division that once shielded the so-called professional sphere' (Bell, 2018, p. 29). Indeed, there have been cases in the past 10 to 15 years where musicians working remotely have developed sufficient skills to create commercial quality outputs. But on the whole, the recording industry preserved a mantra: *these recordings were ok for demos, but we'd like to you come into the studio when we do the real thing.*

Clear and undeniable success from renowned DIYers such as Grammy award-winning producer Finneas Baird O'Connell, older sibling and coproducer to the highly acclaimed Billie Eilish, has gone some way to changing this view, but it has been the COVID-19 pandemic that rocked the world in 2020 that has shifted the goalposts. Out of necessity due to local lockdowns, cooperation and collaboration have dramatically moved into the online virtual space, an adjustment that is challenging many of the traditional notions of recording studio coproduction. In Australia, it is evident that this significant change is not reverting anytime soon, and as this book will show, the most powerful gains to come out of this, *convenience and cost-effectiveness*, are so far, driving a new wave of coproduction for session singers in this country.

Defining coproduction

Coproduction is the work or labour that is performed to create a productive outcome. In relation to recording studio production, outcomes are usually musical artefacts or recordings. Coproduction may involve a collection of participants assembled in one place, but it can also include others who are not present (Sawyer, 2012), whose influence is gathered and disseminated amongst the group or, perhaps, even inferred. For two or more people to coproduce something requires promotive relationships that set out to achieve a common goal (Johnson & Johnson, 2009).

Promotive relationships also require participants to inhibit *competitive* behaviours (Johnson & Johnson, 2009), which can impede the coproduction process and create outcomes that may only be desirable to one or some of the participants. In contrast, *cooperation* may be understood as an assemblage of roles and responsibilities between a collection of participants, that aims to achieve a predetermined goal (Dillenbourg, Baker, Blaye, & O'Malley, 1996). In the recording studio, which has a long-standing history of predetermined roles and responsibilities, cooperation is the starting point from which effective collaboration can emerge.

Collaboration 'requires phases of delineated cooperation', in which the collaborators 'work separately and in parallel' within their particular roles and responsibilities (Rowe, 2020, p. 795). In the recording studio, it is theoretically possible that coproduction could simply be one long phase of delineated cooperation, in which all of the participants engage in their own particular roles and responsibilities, and in which power hierarchies determine the flow of communication. However, participants often seek promotive relationships that demand 'a type of communication amongst the participants that can expand the conceptual approach to the shared task' (Rowe, 2020, p. 795), that is, in collaboration, there is often a willingness to suggest things and have them scrutinised by others for the sake of new emergent ideas or even unintended outcomes (Raelin, 2006).

However, because of the traditional organisational structures that exist in recording studio production, it cannot be assumed that there will be an even distribution of access to deliberations and value within those discussions (Dillenbourg, 1999, p. 7). Instead, each collaborator must find their *rightful place* in the coproduction, constantly socially regulating (Hesse et al., 2015) in order to not only remain included in the creative process but also maintain the required distribution of leadership (Glew, O'Learly-Kelley, Friggin, & Van Fleet, 1995).

It could be argued that effective collaborators can recognise and fluidly adapt to the distinct needs of different collective decision-making moments. *Collaborative dexterity* might therefore be understood as a complex phenomenon, involving continual responsibilities and choices from all involved, to stop collaboration slipping into distinct forms of social interactions, like cooperation or competition.

(Rowe, 2020, p. 796)

In this book, you will witness the collaborative dexterity session singers perform in order to adapt to various environments and decision-making moments.

Defining the vocal booth

This section describes vocal booth performance in terms of the location, materials and performance roles that are central to the practice. This is followed by a short autobiographical description of the author's personal experience as a recording session vocalist within the Australian popular music industry.

Vocal booth is a term used to describe an acoustically treated space that can be used to record vocal performances in a controlled and, in some cases, isolated sonic environment. This control of the sonic environment enables the recorded sound to represent more closely the tonal quality required or desired by the producer and recording engineer. Two main items of technology feature in any vocal booth: the microphone, which is used to record a singer's performance; and headphones, through which a singer can hear the accompanying musical track as well as their own voice as it is recorded. In cases in which the vocal booth isolates the singer from other studio personnel, these two technologies are also used to communicate between rooms and recording spaces.

For the purposes of this book, a vocal booth is defined as a recording space equipped with these two functioning items and operating in conjunction with further technologies that facilitate recording, simultaneous playback and multi-tracking.[1] The term is not limited to confined spaces, nor professional recording studios, but rather it hinges on the function of the space to enable collaboration between studio personnel, technology and a vocal performance to result in a recorded artefact.

The vocal booth is the site of creation for most recorded vocal music heard today. The development of recording technology and its integration into virtually every field of music performance means that many, if not most, singers will, at some point in their careers, experience performing in a vocal booth.

I have specifically chosen to focus on the *session singer* in this publication as it brings a conformity of social and cultural structure to the analyses and provides a launching place from which this research area may expand in the future. Interviews and observations with participants are coupled with the author's first-person account, enabling the reader to grasp a nuanced experience of vocal booth performance in various situations and contexts.

In comparison, *recording artists* use the vocal booth to produce works on behalf of themselves and/or a company to which they are contracted. In such situations, engineers and producers are hired to facilitate the recording process. An artist in the vocal booth creates a very different cultural paradigm and one that is not within the scope of this book. Artists hire producers, which is the more commonly held conception of how singers and producers interact. I am interested in what happens when the producer hires the singer, and what it means when the singer is the one getting paid.

A vocal recording session is generally made up of two parts. The pre-recording phase familiarises the participants with the aims, methods, materials and technologies of the session and orients them to the task. The participants may negotiate their preferred strategies for the recording process, discuss references to existing works and distribute materials such as scores, charts and lyric sheets. This event usually occurs in the control room of the recording studio.

The recording phase is the process of capturing individual performances, or takes, through the use of recording technologies. In popular music, the vocal booth is not simply a recital venue used to record an existing work, but rather a production site in which the plans are not entirely clear from the outset (Greig, 2009), where studio personnel collaborate to craft several sonic variations of a work.

Recording begins with the first take, which is usually followed by reflection on that performance by the participants. This reflection may be discursive and involve the studio personnel focusing on ways to vary and improve the performance on the next take, or it may simply involve repeating the action in the hope of producing a better outcome. Session singers demonstrate expertise through their control of subtle variations in each successive take and are often heavily invested creatively, continually doing, reflecting, improving, transgressing and discovering what makes the *perfect* take. Sometimes, they will share their own ideas, highlighting the often-blurred line between collaboration and authorship (Campelo, 2015). However, outcomes might also be completely unexpected or even contradictory to the plans of the performer and producer. The quality of each performance is often judged 'not simply "correct", but special, idiosyncratic, individual' (Williams, 2010, p. 60) by anyone (or all) of the studio personnel.

The vocal booth performer is therefore an important cultural intermediary, yet the role of this participant has remained relatively unstudied at the academic level. Literature in the field of recording studio production has been mostly written from the perspective of engineers and producers, which belies the significance of the vocal booth performer as a topic of research. A substantial amount of non-academic literature such as trade magazines, books, online interviews and blogs offers singing tips for recording practice and engineering tips for recording vocals (Beck, 2002; Crich, 2004a; Crich, 2004b; Jackson, 2004) and the role of the studio session singer has entered the public domain through feature documentary films such as *Twenty feet from Stardom* (Friesen & Neville, 2013). However, academic literature has mainly focused on aspects of genre comparisons, back-up singing, gender and race (Fast, 2009; Faulkner, 1971; Greig, 2009; Williams, 2009; Williams, 2010), and none of these publications have dealt in any depth with coproduction from the perspective of the singer.

This book aims to reveal the nature of the session singer as an *immediate cultural intermediary* who has the potential to affect an outcome directly (McIntyre, 2008a). Finding out the answer to the question 'what do session singers do?' will further challenge notions of how coproduction in the recording studio is structured and what assumptions and expectations need to be re-examined.

About the author

My commercial career began at 15 years of age, playing and singing Billy Joel and Elton John ballads on an upright piano in a Northbridge cafe in Perth, Western Australia. I played for 4 hours for 20 dollars, a commercial transaction that placed a value on my services as a performer for the first time. Apart from my fee, this transaction also included a memorable set of performance parameters – 'don't sing too loud, stick to the original melody and try to sound as much like Billy and Elton as you can'.

Since that time, most of my commercial performance opportunities have been in cover bands, backing vocal sections and recording studio work. Many of these situations presented similar performance parameters, which developed my skills as a session singer and elevated my standing within the Australian music industry. I have worked as a live and recording backing vocalist for internationally famous artists such as Barry Gibb (from the Bee Gees), Olivia Newton-John, John Farnham and Tina Arena, while additional freelance session recording work has included advertising jingles, song demos, commercial television house bands and guest features.

For over 25 years, I have developed a reputation as one of Australia's leading session singers and procured a broad tacit knowledge of vocal booth performance. This has resulted in close relationships with Australia's leading artists, producers, engineers and music directors and familiarity with evolving studio technologies. My expertise and knowledge within the field of vocal booth performance enable this study to produce an explicit and comprehensive understanding of this specialised practice from an insider's perspective.

Method

In this book, I examine the responses to semistructured qualitative interviews of 20 participants, all of whom are professional session singers within the Australian music industry. The author, who has also sustained a session singing career for over 25 years, provides autoethnographic case studies that amplify the findings and open a window to the practices of session singing as they existed in 2019, before the COVID-19 worldwide pandemic.

The research design of these interviews is reflected in the order of the upcoming chapters in this book, beginning with participants' first impressions of studio recording, and how specific skillsets were developed that led to session singing becoming a large part of their careers. Central to this are the socio-cultural aspects of power dynamics, communication and empathy. The next chapter investigates the effect of singers becoming creative collaborators in the coproduction process and how this can both enhance and compromise creative performance. Chapter 4 looks at the various *technical* environments that can affect performance and how the downsizing of recording studios in the second wave of recording production changed the role of the session singer. This is followed by an inquiry into the third wave of remote recording, which has placed even more responsibility in the hands of session singers and raised many new questions about contemporary coproduction. In the final chapter, I summarise the findings and offer suggestions for the future of coproduction in the recording studio.

Note

1 Multi-tracking describes the act of recording multiple performances with the facility to play back any or all of these tracks while recording further performances.

References

Beck, D. (2002). *The musician's guide to recording vocals*. Milwaukee, WI: Musicians Institute Press.

Bell, A. P. (2018). A history of DIY recording. In *Dawn of the DAW: The studio as a musical instrument* (pp. 1–30). New York: Oxford University Press.

Bourdieu, P. (1990). *The logic of practice* (R. Nice, Trans.). Stanford, CA: Stanford University Press.

Burgess, R. J. (2014). *The history of music production*. New York: Oxford University Press.

Campelo, I. (2015). "That extra thing": The role of session musicians in the recording industry. *Journal on the Art of Record Production, 10* (July 2015). Retrieved from http://arpjournal.com/asarpwp/that-extra-thing-the-role-of-session-musicians-in-the-recording-industry/

Cole, S. J. (2011). The prosumer and the project studio: The battle for distinction in the field of music recording. *Sociology, 45*(3), 447–463.

Commonwealth of Australia. (1998). *Copyright Amendment Act (No. 2) 1998,* AGPS, Canberra.

Crich, T. (2004a). Recording vocals. *Professional Sound, 15*(5), 56. Retrieved from https://search.proquest.com/docview/229488614?accountid=12528

Crich, T. (2004b). Recording vocals. *Canadian Musician, 26*(2), 75. Retrieved from https://search.proquest.com/docview/216521094?accountid=12528

Davis, R., & Parker, S. (2013). *More than microphoning: Capturing the role of the recording engineer from the 1980s to the 1990s.* Retrieved from https://api.equi noxpub.com/articles/14846

Dillenbourg, P. (1999). *Collaborative learning: Cognitive and computational approaches.* Oxford: Elsevier.

Dillenbourg, P., Baker, M., Blaye, A., & O'Malley, C. (1996). The evolution of research on collaborative learning. In E. Spada, & P. Reiman (Eds.), *Learning in humans and machines: Towards an interdisciplinary learning science* (pp. 189–211). Oxford: Elsevier.

Driver, C. (2015). The collaborative mix: Heidegger, handlability and praxical knowledge in the recording studio [online]. In *Into the mix people, places, processes: Proceedings of the 2014 IASPM-ANZ conference.* Dunedin, NZ: International Association for the Study of Popular Music, IASPM Australia-New Zealand Branch Conference.

Fast, S. (2009). Genre, subjectivity and back-up singing in rock music. In D. Scott (Ed.), *The Ashgate research companion to popular musicology* (pp. 171–188). Abingdon, UK: Routledge.

Faulkner, R. (1971). *Hollywood studio musicians: Their work and careers in the recording industry.* Chicago, IL: Aldine-Atherton.

Friesen, G. (Producer), & Neville, M. (Director). (2013). *20 feet from stardom.* Los Angeles, CA: Gil Friesen Productions.

Gibson, C. (2005). Recording studios: Relational spaces of creativity in the city. *Built Environment, 31*(3), 192–207. Retrieved from www.ingentaconnect.com/ content/alex/benv/2005/00000031/00000003/art00003?crawler=true

Greig, D. (2009). Performing for (and against) the microphone. In N. Cook, E. Clarke, D. Leech-Wilkinson, & J. Rink (Eds.), *The Cambridge companion to recorded music* (pp. 16–29). Cambridge: Cambridge University Press.

Glew, D. J., O'Learly-Kelley, A. M., Friggin, R. W., & Van Fleet, D. D. (1995). Participation in organizations: A preview of the issues and proposed framework for future analysis. *Journal of Management, 21*, 395–421.

Hesse, F., Care, E., Buder, J., Sassenberg, K., & Griffin, P. (2015). A framework for teachable collaborative problem solving skills. In P. Griffin, & E. Care (Eds.), *Assessment and teaching of 21st century skills* (pp. 37–56). Dordrecht: Springer.

Horning, S. S. (2004). Engineering the performance: Recording engineers, tacit knowledge and the art or controlling sound. *Social Studies of Science, 34*(5), 703–731.

Howlett, M. (2012). The record producer as nexus. *Journal on the Art of Record Production, 6* (June 2012). Retrieved from http://arpjournal.com/asarpwp/the-record-producer-as-nexus/

Jackson, B. (2004). Recording vocals. *Mix, 28*(10), 32. Retrieved from https://search.proquest.com/docview/196873739?accountid=12528

Johnson, D. W., & Johnson, R. T. (2009). An educational psychology success story: Social interdependence theory and cooperative learning. *Educational Researcher, 38*(5), 365–379.

Krueger, A. B. (2019). Streaming is changing everything. In *Rockonomics : A backstage tour of what the music industry can teach us about economics and life* (pp. 177–204). New York: Currency.

Leyshon, A. (2009). The software slump?: Digital music, the democratisation of technology, and the decline of the recording studio sector within the musical economy. *Environment and Planning A, 41*(6), 1309–1331.

McIntyre, P. (2008a). The systems model of creativity: Analyzing the distribution of power in the studio. *Journal on the Art of Record Production, 3* (November 2008). Retrieved from www.arpjournal.com/asarpwp/the-systems-model-of-creativity-analyzing-the-distribution-of-power-in-the-studio/

Mellor, D. (1996). *How to become a record producer, Part 1: What is a record producer?* Retrieved from https://web.archive.org/web/20150607013722/www.soundonsound.com/sos/1996_articles/jan96/recordproducer1.html

Moorefield, V. (2005). *The producer as composer.* Cambridge, MA: MIT Press.

Papadopoulos, T. (2000). Copyright, Parallel Imports and National Welfare: The Australian Market for Sound Recordings. *Australian Economic Review, 33*(4), 337–348.

Raelin, J. (2006). Does action learning promote collaborative leadership? *Academy of Management Learning & Education, 5*(2), 152–168.

Rowe, N. (2020). The great neoliberal hijack of collaboration: A critical history of group-based learning in tertiary education. *Higher Education Research & Development, 39*(4), 792–805. doi: 10.1080/07294360.2019.1693518

Sawyer, R. K. (2012). *Explaining creativity: The science of human innovation* (2nd ed.). Oxford: Oxford University Press.

Théberge, P. (2004). The network studio: Historical and technological paths to a new deal in music making. *Social Studies in Science, 34*(5), 759–781.

Watson, A. (2014). *Cultural production in and beyond the recording studio* (Vol. 47). London: Taylor & Francis Group.

Williams, A. (2009). "I'm not hearing what you're hearing": The conflict and connection of headphone mixes and multiple audioscapes. *Journal on the Art of Record Production, 4* (October 2009). Retrieved from http://arpjournal.com/asarpwp/"i'm-not-hearing-what-you're-hearing"-the-conflict-and-connection-of-headphone-mixes-and-multiple-audioscapes/

Williams, A. (2010). Navigating proximities: The creative identity of the hired musician. *MEIEA Journal, 10*(1), 59–76.

2 Stepping into the sociocultural space

First impressions

For any musician, stepping into a recording studio environment for the first time can be a memorable experience. Minds are enveloped with new sensations, smells, sights and sounds that remain with them for a lifetime.

> The first time I'd been in the studio as a singer was when I was really young in England. I remember walking into the BBC and the smell of the wood in the recording studio. I felt immediately at home and thought that's where I want to be. And I've gravitated back towards that since then.
>
> (SA)

Many of these experiences occurred at a relatively young age, when there are often fewer inhibitions and singers are able to simply soak up the atmosphere and let it 'change their world'.

> My very first studio experience was when I was 14. It was amazing, because I had never been inside a real studio before. As a young girl I had this amazing experience of the cans, and the microphone, and the piano. I remember feeling really comfortable and loving the experience and being really excited by it and just kind of taking it all in my stride. I think I just belonged. I felt belonging instantly.
>
> (JM)

> I was maybe 15 or 16 doing guest vocals on like a power metal group. I was singing along with the guy who was singing lead and had to put myself in there and just sort of fit in – just do what I had to do for them to be happy with the outcome. I think I enjoyed adapting, I enjoyed being a part of something that just wasn't an everyday thing. And then

DOI: 10.4324/9781003146865-2

I think I started to understand how recording can take you into these really different worlds. And I enjoyed that.

(SG)

I did my first session at a studio in Sydney. It was run by a legendary engineer called John Frohlich. We were working to tape. There wasn't any Pro Tools. There weren't any computers. I'm gonna say it was 1988 so I was only about 15. I just had to do one little tagline at the end, but I can just remember still, the feeling of the pressure, and the adrenaline that was rushing through was different. I'd already been doing gigs, and this was a different kind of feeling. And I liked it. I liked that I had to do what I feel that I'm here to do. And I knew that very young, enjoyed the feeling of it. And I just knew that it was something that really came naturally to me.

(DP)

Reports of *feelings of belonging* early on are very strong amongst established session singers. It may be a connection to the environment, but it may also be a recognition that this workspace and the opportunities it affords suits a particular personality or *identity* with a particular skillset. Without necessarily thinking about it, a decision is often made at a very young age to cultivate a relationship with *recording studio people and places*.

I'd been sending tapes to local studios as you did in those days. One producer liked the sound of my voice and booked me for jingle that he was doing at the time. It went to air and I was able then to use that as part of my CV to go and speak to other jingle writers, producers and the like. It was about 1978 or 79, and I was 16 or 17, and it kind of kicked off from there. I ended up gathering what's turned into now over 1000 ads. Which is amazing. Yeah, that's how much work there was in those days.

(SW)

The number of recording sessions during this 'golden era' of session work established networks of musicians as well as the cultural and social standards that others would follow.

I did an American Express commercial on a friend's recommendation, and I thank him from the bottom of my socks for that, because it was just the beginning of so many things. Because when I did that, it just spread out to the world. And I could do more sessions for people. And during that time during the 80s, it was just huge. I was doing five

sessions a day after that and running between studios and making
quite a good amount of money too.

(SA)

By the 1990s, the proliferation of home recording equipment precipitated an
increase in the number of private recording studios. Some of these maintained
an 'industrial' veneer by situating in central business locations, while others
reverted to suburban private dwellings. Singers who emerged in this period had
an experience formed by a hybrid mix of some of the old and some of the new.

> I started to get a few sessions, just doing little ads and stuff like
> that. And they were more in like smaller studios, production stu-
> dios. And it would always just be me and one producer. And they
> usually happen super-fast. You're in and out within, you know,
> 45 minutes, just singing really simple jingles. I was young and
> I've always had a kind of happy sounding voice. So, I always got
> booked specifically for something kind of youthful and bright and
> chirpy. They were always the gigs that I got booked for.
>
> (JMc)

> I got into jingles in Sydney, I reckon it was about 2005. Back then
> it was definitely booths. Now when I go and do jingles, they're
> much more like home setups, than they are booths.
>
> (CS)

Fortunately for singers who began their careers at this time, there were still
producers, engineers and musicians who had come through the first wave
of recording studio production that demanded high standards of social, cultural and musical excellence. Receiving direction from these experienced
people, albeit often in a more downsized environment, meant that much of
that experience was passed on.

> Richard Lush was the engineer, who famously worked with the
> Beatles in Abbey Road when he was a young guy. And Richard
> was amazing. So, talk about, you know, cutting your teeth with the
> best of them. Richard taught me a lot, but I also loved how intui-
> tive he was, and he knew what you needed.
>
> (GC)

> I started learning singing from an amazing man called Alan Dean.
> He'd had an incredible career in England, then went to America,
> worked with all the greats, came out to Australia on a tour and

never left. And he started working in advertising, and producing music, jingles, and working doing corporate stuff. He took me under his wing and started to take me to sessions. So, my mum graciously would write me a note and I would go with Alan. And then he actually got me to sing on my first ever session.

(DP)

Working alongside more established session musicians left a lasting impression on up-and-coming session singers. Learning studio conventions and etiquettes from an early stage enhance an inexperienced singer's opportunities to be re-hired and form a good reputation amongst a wider network of studio personnel.

One of my very first sessions was with Susie (Ahern). And I remember I knew who Susie was. And I was in such awe, and I just, I kind of watched her secretly in the session, how she held herself and how she did everything and learned by watching her.

(JM)

I'm sure you know, Danny (DeAndrea) and Darren (Percival). And the two of them would do a lot of sessions and I had sort of known them since I was a teenager and then they would often get me in as a third singer. So I learned a lot from those guys, both in a live setting but especially in the studio. They were obviously you know, consummate professionals and I really didn't know much about what I was doing at that point. So it was really, really nice to have that. Some of us have had that mentoring, which was, you know, for me was it was a really big part in being able to solidify what I was doing as a session singer fairly quickly, learning quickly, so that I made sure I got called next time.

(JMc)

Growing up in these situations, I was lucky to work with the generation before me who were ultimate professionals, and so you learn when to shut up, when to listen, when to just say yes, when to offer some help or some advice. And that's all relational. You have got to read the room yourself.

(GC)

Reading the room

In many of the interviews conducted for this book, participants expressed the importance of *reading the room* in any coproduction event. Identifying cues, both verbal and non-verbal, provides a singer with information about the socio-cultural structures of a particular recording session. More specifically, what the roles and power structures are within the group of participants, and how communication is likely to be conducted. It appears to be one of the first social skills a session singer learns, and one they take with them into every subsequent recording experience. During the process of reading the room, a session singer develops, either consciously or subconsciously, a level of *emotive trust* ('based on one's personal feeling about another') in accordance with the social dynamics, and *capacity trust* (based on one's judgement about another's capacity for competent performance in a workplace) in accordance with the cultural dynamics (Ettlinger, 2003, p. 146).

> It's such an important thing, reading the room, and knowing how to proceed depending on whether someone's a spiky character or not, you know. And it's important that you know how to walk through a situation and do the best job that you can possibly do.
>
> (SA)

> You can feel people's emotional energies immediately as you come into a situation. I know where they're at within a minute, within a glance. And I will either tread carefully with that person, or embrace them, or be able to circumnavigate them, to achieve what we need to achieve in the time that we need to achieve it.
>
> (GP)

> I reckon you work it out pretty quickly, if you're working with somebody for the first time. I mean, I always start by listening really hard. And then just gauging as we kind of get to know each other in the process.
>
> (JM)

> I always try to gather as much Intel as I can, as quickly as I can, especially when I'm working with someone new. There's such a great varying degree of different skill sets that you might encounter in the studio.
>
> (JMc)

Going into any situation expecting to not know what the mood or operation will be within an organisational system is a daunting task, but one that session singers have to do most of the time. Singers must not only adapt their voice, musicality and style to suit the task, they must also adapt their behaviour to the particular environment, which means that session singers are continually vigilant to social and cultural cues. For some singers, *reading the room* starts before the session, sometimes even before agreeing to terms. Preparation can include obtaining as much (or as little) information as one is prepared to ask for.

I like information. So, I always make sure I know quite a lot before I get there. What will I be doing? Who am I going to be with? Who's booked on the session? And that's really important, because the dynamics and the energies and stuff. I got told early on *effortlessness takes preparation*. I've never forgotten it. I stand by it daily. So, I do a lot of work before I get there, kind of get an idea of the style. What tonally might be happening. It's just quick questions, you know, send an email or whatever.

(DP)

I kind of have a set little idea of what I ask now when somebody rings me to book me for something. And they're usually questions like *how long? what's it for?* You know, *what's the fee? Where's it going to be used?* all that kind of thing.

(MS)

Prior to the session, because I'm catholic, I would ask them, *does this lyric have anything that is going to compromise what I believe in? Because if it does, I have to give away the session.* So, we go in prepared for that.

(GP)

A strong understanding of these cultural cues and a flexible social mindset going into any session appear to be hallmarks of *professionalism* in session singing, and the way session singers accumulate cultural and social capital (Bourdieu, 1990). But understanding alone does not guarantee a successful outcome. Commitment to the goals that have been set by the producer and common attachment to 'governing variables' (Argyris & Schoen, 1974), or baseline assumptions, beliefs and feelings about the way coproduction might occur, facilitate promotive relationships and build a collaborative and cooperative environment.

There is an ownership that I adopt when I walk into a studio to the project. I make it personal. So whether I realise I am or not, I'm thinking about it in that way. Every job is personal, and I want to make sure that I'm doing everything I can to give you what you need.

(GC)

Professionalism isn't just about the work we do vocally. I think it's equally balanced with all that other stuff – energy, how we present ourselves, being on time, all the financial business stuff as well. It's a big part of it. So yeah, just a lot of organisation. I want someone to book me again so I am pretty conscious of how my behaviour might affect this.

(DP)

I try and understand what they're wanting me to do. I really just try and understand them as a person and I try to read their personality. Because if you're quite a sensitive person, I think things can get taken the wrong way. I think every producer, every engineer, I just expect them to be different. And I expect everything, the environment, their nature, I expect it to all be different. I like the challenge of adapting and trying to decipher for myself and trying to just get through the session. At the end of the day, you want that exact same result that they do.

(SG)

There is an overwhelming sense that session singers find *their place and their people* in the recording studio. But it is also clear that many initial forays into this environment exhibit a certain amount of anxiety and, at times, naivety around expectations and assumptions. Most singers admitted that when they started out, they did not really know what they were doing and relied on the experience of other participants around them to show or tell them how coproduction should occur. However, every recording session, regardless of how experienced you are, has the potential to present an unknown mix of roles and hierarchies that require careful navigation.

Power dynamics

An assumption that can be invoked in almost every recording session is that the producer is at the top of the chain of command. Whether or not they hold the economic capital, they are still the individual a session singer is working for, and potentially, if everything goes well, will work for again. This could be termed the *chain of command* assumption. An understanding of the traditional roles and power structures in the recording studio enables a relatively inexperienced session singer to aptly defer all communication and decision-making to the producer, whose role it is to filter communications and direct everyone towards a common and overarching objective.

I think, if you looked at it as a hierarchy, as a pyramid, perhaps your composer slash producer has to be at the top. Because really, they're the one that's driving the bus, the engineer is under their instruction. And the clients are somewhere just below them, I guess, because it's still their vision that the producer is interpreting and trying to construct in a musical way. It's an interesting dance, I find, and I think a good producer has to be such a good people person and a good translator, between agency, the client, the singer and the end product, because sometimes the client, whoever they are, sometimes they're not musical. So, a composer/producer has got to listen and decipher what it is they're trying to say, and then relay that to the singer. It's a job of interpretation as much as it is of being a musician, you know, so I kind of find that fascinating.

(GC)

The *chain of command* assumption can be challenged in a number of ways. Firstly, when the party holding the economic capital (e.g., client) chooses to give direction to the producer, it can have a dramatic effect on the coproduction process from the perspective of the vocal booth. There is also a strong impression from the session singers interviewed that coproduction works most efficiently when the triumvirate of producer, engineer and singer are occupied by people in well-defined roles. In fact, the mere presence of another person in the room can drastically change the structures of communication.

I think when it's just, the singers, and the producer and the engineer, we can work faster and there's also a lot more to-and-fro that's creative and productive. And when the client is in the room, that's when we have to be quiet. There's no kind of to-and-fro anymore, it's between the client and the producer. You kind of learn to be quiet and when to offer any kind of suggestion.

(JM)

When it's just me and another producer, it's very, very free. Like I can say whatever I want. And you know, say all my ideas and stuff, but when it's a big team like that, you kind of just have to listen.

(JD)

Another common way the *chain of command* assumption can be challenged is when the power distribution, which may be notionally measured as relative capital (cultural, social and symbolic), shifts towards the session singer. While the traditional structure could be considered *high-low* (producer-singer), other power distributions could be described as *even* (mid-mid) or *reversed* (low-high). These alternate power dynamics can impact greatly on the coproduction process.

I'm really trying to get a gauge of everybody's personality, everybody's skill set. I have great confidence in my skill set and I think I'm also able to be very flexible. I've worked with some people who absolutely know exactly what they want and they're incredible, much more experienced than I am, and I'll just allow myself to be directed by them. But it's a really broad spectrum. You know, sometimes I've worked with people who really are not as equipped to convey what's best for the end result. If I'm starting to get a feeling that maybe, you know, there's some gaps, I'll start listening in a different way, preparing myself that I may need to contribute, as opposed to just being directed. I'm always open to that and often this has to happen really quickly, and you have to kind of make that choice of *what exactly is my role going to be? How do I work this situation best?*

(JMc)

I stand by the fact that experience is currency, and our odometer of time standing at that microphone means something. It is worth something. And if you have an odometer with a big number, you're in the position to be able to help that environment by sharing what needs to happen in order for the job to get done. Because otherwise no one's taking charge. Especially when things aren't flowing, I'll just go in there, sit down and have a chat. *Yeah, look, I'm really excited about getting this vocal up. But under the current energy and situation, I'm not gonna be able to do that. So you guys need to work out how you can get this to happen, because you've got a vision and you don't agree with it.* You know, so you got to be the boss.

(DP)

It's such a fine line between doing your job to the best of your ability to aid the session and stepping on people's toes.

(GC)

You still want them to be happy with the result, even if you know that they're not giving you much direction. So I guess you just sort of try to bring your experience and, you know, whatever intelligence you've got about what you're trying to achieve. You're doing all this extra work to make it seem effortless, and to feel comfortable for the person who perhaps should be taking on that role a little bit more. You subconsciously take on these roles, but in a way where you want them to still feel like they're in control.

(SG)

Emotional labour

Conscious effort to make others feel *comfortable* in the recording studio environment is a concept that has been discussed in regard to engineers and producers. Watson and Ward (2013) state that 'Recording studios are *emotional spaces* in which producers and engineers regulate their own emotions by way of managing the emotions of musicians and recording artists as they perform' (p. 2905). Musical performance is an 'intimate' encounter, in which emotions are routinely heightened, as performers share personal experiences with their audiences (Anderson & Smith, 2001, p. 3; Wood, Duffy, & Smith, 2007). Recording studios are also social spaces where the emotions can be further intensified by the nature of the environment and the human collaboration occurring therein. Support and encouragement are assets that producers and engineers use to facilitate the creative process (Leyshon, 2009, p. 1316), performing *emotional labour* (Hochschild, 1983) to control the vibe in the studio, relax the musician or recording artist, and to get them in the right frame of mind for their performance (Watson & Ward, 2013). This requires a combination of empathetic emotional labour towards the performer (Korczynski, 2009) and 'professional' emotional neutrality that suppresses reactions that could induce a negative emotional experience for the client (Smith & Kleinman, 1989). Missing from the literature is an account of the emotional labour performed by the recording musician in this setting, and questions might be asked about the extent to which this type of performance is imperative to a successful recording session.

Session singers interviewed regularly spoke of *empathy*, a willingness to put oneself in another's position in order to understand what they need, and how their situation can be helped in order to produce the best outcome. This is a vital part of collaboration and coproduction that is not only performed by participants in the control room, but also by the session musicians.

> Session singers are extraordinary singers. They're not just singers who can hold a tune. They can allow themselves to be a component of a greater voice. And to do that, you have to put your ego aside, you have to put your identity aside and just recognise where your strengths are, how to harness those and how to combine those with people who have different strengths, to make something that's really whole and beautiful. And to do that, you have to be able to listen, you have to be able to feel, you have to be able to empathise, you know, you have to be a fine human being in a way to do that, because that's what fine human beings do.
>
> (LF)

> I feel like the vibe is everything. It starts when they greet you at the door. Because, yeah, if you don't get like a warmth from someone it can make you very vulnerable. I guess it does have to go both ways. If you don't feel like you're getting that from someone, you do probably have to find ways to connect with them on some level,

even if it's like asking for background on the song. *What's this project for?* Or, you know, *where did it come from?* or *How did you get this job?* Just to start some kind of conversation. You've got to set a rapport with people, otherwise you can hear it in the recording.

(JC)

If someone's being very fronty, but they may not be great at what they do in a producer capacity, I guess the best thing I can do is just reign myself in and go, *Alright, cool, I'll ask them questions.* If I'm trying to help us collectively get the best out of the situation, I'll ask them questions. I'll dial back the laughter and be more business about it, I guess, but you still got to keep an element of lightness in the room somehow. Otherwise, it can all turn really bad. They might be driving the bus, but I'm the centre of attention at that point, because we're there to do my job and to get my job right. So invariably, even if it's a negative situation, I'll try and keep a bit of friendliness between the two panes of glass.

(GC)

Sometimes the empathy becomes more strategic and takes on a form of *conditioning*. Similar to the way a producer or engineer will celebrate a performance by the singer, even if it was not the perfect take, session singers are aware of the need to sometimes condition their coproducers.

There's definitely conditioning going on. Working in real close proximities, it's quite intense, and actually quite intimate, very personal. So you are trying to make it as best a situation as you can, as comfortable for yourself. Sometimes conditioning that other person just makes it more comfortable for you. If we're not comfortable with something, sometimes we'll just say it. But a lot of the time it is about working around personalities and keeping somebody happy. It's just another element, this hidden element that's in the room, that you are working around constantly. I guess tactics are always, you know, asking what they think. Like, *here's a suggestion, what do you think?* always putting it back on them. And, you know, just making suggestions, as opposed to telling them what you think actually should happen here.

(SG)

And so a lot of that is about management skills, people management . . . with some of the producers, it's about managing them . . .

the way that you present is, *is this is what you mean?*, even though, we've already worked it out, you know?

(LF)

We massage their ego so often, you know what? That's where it's like a normal job sometimes.

(AL)

I think, again, it becomes about your situational awareness of understanding the different personalities in the room – who's dominant, who's not, I think I'm quite good at playing either role. So I'm happy to step in and go, *Okay, let's do it like this.* I'm also happy if there's someone else there who I feel needs to be like that. And some people do. Some people can only be like that, I'm also happy to defer to them and say, *Okay, I'm just here for the ride.* But it can be challenging when you have people with vested interests in the room with you who maybe don't under-stand the work process. That can be uncomfortable, and I think has to be, you know, massaged a certain way. Otherwise it can become problematic.

(JMc)

Summary

Recording studios are socio-cultural spaces that session singers must *assess* each time they enter a new workspace. *Reading the room* entails forming an understanding of the roles and responsibilities of each participant, and judging their social, cultural, symbolic and economic capital. From this assessment, a session singer forms a set of *governing variables* or baseline assumptions, beliefs and feelings about the way coproduction might occur. Assessment continues throughout the recording event, with many inter-viewees in this study describing their ongoing attempts to empathetically understand and share the feelings, beliefs and assumptions of other partici-pants in order to achieve a successful outcome.

As a recording session progresses, levels of trust and empathy may vary, which can in turn alter the way roles and responsibilities are actioned in the coproduction process. High levels of emotive and capacity trust within a coproduction event enables participants to openly share their feelings, beliefs and assumptions, which facilitates the accomplishment of a shared goal. Regardless of the relative power distribution, trust and empathy sustain communication and a continual collective *updating* of the shared governing variables. When these levels of trust diminish, participants will often privately recalibrate their investment in the coproduction.

As the socio-cultural environment changes, session singers may choose to focus solely on their role as the singer, aiming to just *get the job done* despite the conditions. Conversely, upon recognising that they have the expertise to fulfil a limitation in the coproduction process, a session singer might choose to adopt additional roles and step into the *creative space* that is traditionally the domain of the producer, songwriter and engineer. In so doing, a session singer becomes more than *just a vocalist* in the coproduction process. The following chapter reveals from the vocal booth perspective, how such movements can happen, how movements in trust, roles and responsibilities affect session singers, where the *line* should be drawn in particular instances, and the effect of unforeseen responsibilities on performance and coproduction.

NOTE: See Appendix 1 to read an autoethnographic case study that demonstrates the role empathy can play in supporting trust, regulating emotions, interpreting directions and shaping a session singer's approach to the task.

References

Anderson, K., & Smith, S. J. (2001). Emotional geographies. *Transactions of the Institute of British Geographers, 26*(1), 7–10. Retrieved from www.westernsyd ney.edu.au/__data/assets/pdf_file/0008/150947/Anderson_and_Smith_Emotional Geographies_ICS_Pre-Print_Final.pdf

Argyris, C., & Schön, D. A. (1974). *Theory in practice: Increasing professional effectiveness.* San Francisco, CA: Jossey-Bass. http://doi.apa.org/psycinfo/1975-03166-000

Bourdieu, P. (1990). *The logic of practice* (R. Nice, Trans.). Stanford, CA: Stanford University Press.

Ettlinger, N. (2003). Cultural economic geography and a relational and microspace approach to trusts, rationalities, networks, and change in collaborative workplaces. *Journal of Economic Geography, 3*(2), 145–171. Retrieved from www.researchgate. net/profile/Nancy_Ettlinger/publication/236869787_Cultural_economic_ geography_and_a_relational_and_microspace_approach_to_trusts_rational ities_networks_and_change_in_collaborative_workplaces/links/0c960 522869816b006000000/Cultural-economic-geography-and-a-relational-and-microspace-approach-to-trusts-rationalities-networks-and-change-in-collab orative-workplaces.pdf

Hochschild, A. R. (1983). *The managed heart: Commercialization of human feeling.* Berkeley, CA: University of California Press.

Korczynski, M. (2009). The mystery customer: Continuing absences in the sociology of service work. *Sociology, 43*(5), 952–967. Retrieved from www.research gate.net/profile/Marek_Korczynski/publication/249825959_The_Mystery_ Customer_Continuing_Absences_in_the_Sociology_of_Service_Work/ links/55ba19b808aec0e5f43e815b.pdf

Leyshon, A. (2009). The software slump?: Digital music, the democratisation of technology, and the decline of the recording studio sector within the musical economy. *Environment and Planning A, 41*(6), 1309–1331. Retrieved from https://journals.sagepub.com/doi/pdf/10.1068/a40352?id=a40352

Smith, A. C., & Kleinman, S. (1989). Managing emotions in medical school: Students' contacts with the living and the dead. *Social Psychology Quarterly*, 56–69. Retrieved from www.jstor.org/stable/pdf/2786904.pdf

Watson, A., & Ward, J. (2013). Creating the right "vibe": Emotional labour and musical performance in the recording studio. *Environment and Planning A, 45*(12), 2904–2918. Retrieved from https://dspace.lboro.ac.uk/dspace-jspui/bitstream/2134/21404/1/EPA45-208Emotionallabourintherecordingstudio-FINAL.pdf

Wood, N., Duffy, M., & Smith, S. J. (2007). The art of doing (geographies of) music. *Environment and Planning D: Society and Space, 25*(5), 867–889. Retrieved from www.researchgate.net/profile/Michelle_Duffy3/publication/228631795_The_art_of_doing_geographies_of_music/links/0deec534de99771f1c000000/The-art-of-doing-geographies-of-music.pdf

3 Stepping into the creative space

As discussed in Chapter 1, coproduction in the recording studio can take many forms, ranging from delineated cooperation where participants work towards a defined and shared goal, to collaborative approaches where each participant is active in expanding the creative concept and working towards a shared but unforeseen outcome.

Regardless of where a coproduction event is initiated, contributors must often adapt, either to the unfolding dynamic of the socio-cultural interactions or to the 'distinct needs of different collective decision-making moments' (Rowe, 2020). In general, session singers interviewed for this book were quite open to these movements and open to the unexpected.

However, despite a wide range of tolerances to these socio-cultural movements, all of the singers interviewed described experiences that caused them to question the amount of creative input they were making and how *stepping into the creative space* translated to the role they were fulfilling, the acknowledgement they received, and the remuneration that was commensurate with these *extra services*.

> I guess when session singers get a call, they're expecting to get in there and just sing. But we often get sort of flung these other roles, you know. I mean, the voiceover actor, or the top liner, and making up the melody. But you've also got situations where sometimes, you know, you might be thrust into like a production role as well, making sort of short production decisions.
>
> (SW)

> I don't feel it's my position to work it out. That's not my job. My job is to sing whatever they're asking me to sing. If there's a lot of different ideas flying around, generally, I sit back and wait for them to work out what the concrete solid idea is. And then to relay that back to me for me to sing it.
>
> (MS)

DOI: 10.4324/9781003146865-3

Songwriting

Of all the extra services a session singer might provide in a recording session, by far the most remarked upon was songwriting.

Sometimes there's that little bit of writing that's involved, definitely, just an opinion on where a melody should go, what would work here, which is great. But there's definitely this crossover that happens, where sometimes you walk away, and you've realised you've done a little bit more than what you were expecting to do. And that's actually made quite a big impact on the result. And what does that mean for you? Where does that leave you? It's just like, this personal little thing that you've walked away with, that nobody else will know.

(SG)

Oh, yeah, I've written quite a number of songs in a number of sessions, including TV jingles. But because I was paid as the session singer to come in and sing it, got no APRA credit for it. It's my melody! When they've got something established, and they go, *can you just sing something at the end there?* And that becomes the hook that gets played before every ad break – that's my hook, man! So yeah, it's a bit frustrating.

(JF)

I feel like sometimes my role is a bloody writer in there, particularly with people's originals. More so than jingles and stuff. Oh my god, sometimes with originals, I swear to God, they go *just sing something*. And I go, *what do you mean, just sing something? This is your song*. And then you've walked out of there and written half their melody for their originals that they were getting you to sing on.

(AL)

Yeah, that happens many times, many times. And of course, you can't say, *well, hang on a second, I need a bit of a co-write on this*. I've never done that.

(NN)

It's very hard, that conversation is really difficult, because there are prickly people who would not actually want to admit the fact that what you've just done, is being part of the creative process.

(LM)

That's happened a lot and that is something that does get under my skin. It's happened a lot where a songwriter, or a producer will ask you to come in and demo a song that they've written. And you're not the intended artist, somebody else is, and you turn up and they have, kind of, half a melody. And they have the lyrics written, but it's like, *oh, it goes something like this*. And I'm now writing the melody, based on some sort of vague indication of what you've said, and I feel that's really kind of, it's very cheeky, and I think that's common, I think it comes up a lot. I've just kind of conceded in those situations so many times. And sometimes people do the right thing, and they'll give you like, a small writing cut or something. And then 99% of the time, it seems to be perceived that that's your job as the singer, to flesh it out, bring it to life.

(JMc)

I've done it too, where they've gone *what do you think of this?* or *can you make something up here?* And then you kind of think, Okay, well, I've kind of half become the singer and a little bit songwriter here as well. So what does that mean? You know, does that mean that I'll get a credit? And you probably won't. Or that you'll get paid more or something like that? But you probably won't. So, it's always a tricky situation to be in. But I think a lot of the time, I know for myself, I just kind of do whatever I'm asked, which is, I think, what most singers tend to do.

(MS)

That's when it can become messy and actually quite frustrating for me. Because it's like, so you've hired me, and here I am. But now you want my creative output. And at that time, I didn't know how to get out of it. Sometimes they knew I was a songwriter as well. They would write a verse and a chorus, *but we don't have the second verse. Do you reckon . . . ?* And that's when I had to draw the line. I rang my publisher, I was like, *yo, listen, I'm in this session. I'm supposed to just do a vocal session. Now it's turning into an actual song* and my publisher was like, *yeah, that's cool, you can do that. You just need to get some credit for it* – basically.

(CS)

I've been in situations where they didn't have a bridge or something, or they didn't have a middle 8 in their song. They kind of say, *oh, what do you think?* And I sort of say, *Well, you know, I was thinking yeah it sort of probably needs a little bit of a lift here, or maybe a four-bar thing here*, or, you know, whatever. And then they'd say, *Well, have you got any ideas?* You are just left with that awkward feeling. Is it now that I say, *am I co-writing this with you? Or am I just suggesting things and you're going to use my ideas?* You know, that happens so many times. It's just, well, it's quite, um, quite annoying. What do you say during that moment? How do I get out of this?.

(SW)

Artistic property

Services such as original vocal phrasing, tonal nuance and basic backing vocal arrangements are generally accepted by session singers as part of the job. But despite a willingness to offer these services and to be invested in the project with a high level of professionalism, songwriting is a form or *artistic property* that appears to be valued above a threshold of just being *part of the gig*. Another form of artistic property that session singers value above this threshold is their personal *sound*.

Sometimes you're asked to do that session, because of what your voice is, and what it brings. You know, composers who will get specific singers that they know aren't necessarily the best singers, but they've got a specific sound. And there's a lot in that, that sound is what they're after. So it makes or breaks the track at the end of the day. It's almost like, a brand, you know.

(JC)

That's happened to me so many times. So you want Carmen the 'artist' basically singing your song, selling it, to whomever you're going to pitch it to.

(CS)

I think one of the things that a good session singer brings is, is an array of voice qualities, right? And not everybody can do that. And not everybody thinks about that, either, you know, that is worth a lot. And the sound thing. I mean, that's so unique. There's one person in particular that books me because they love my sound. And I think they're probably one of the few people that actually gets it because of how I get paid.

(JM)

There are moments where, as the session performer, we are also asked to bring our artistic creativity with it, and our originality because that's what they need and want but they're not actually remunerating you for that situation, the budget's not there for whatever reason.

(NF)

Co-option

Such events are clearly frustrating to session singers who will often feel like they have been *co-opted* into a situation that they do not know how to get out of. The point is reached where professionalism and willingness to be invested in a project crosses a line, leaving one to question 'what does this mean?' If this situation arises while the recording session is in progress, it is considered too uncomfortable and possibly unprofessional to question this verbally. Being *co-opted* in a coproduction can quickly drain a singer's capacity and emotive trust in the other participants.

> Sometimes you leave a session, feeling like you've left a big chunk of yourself there, as opposed to just walking in and doing whatever you've been told. That you've artistically contributed to that piece that you've just sung on. And then I guess whoever's booked you for that session, it becomes their project, and you're not involved past that.
>
> (MS)

> They'll like, say to me, *can you make it like, make it your own?* But it's not my own. It's *your* own. You're just hiring me to sing, you know
>
> (CS)

> Sometimes, I guess, that expectation and having to go above and beyond for something that's not really what you wanted to do or expected, can leave you feeling a bit flat. And you don't necessarily feel comfortable doing it because you are pouring a little bit of yourself into this. And it's not necessarily something you want to be pouring yourself into.
>
> (SG)

The gap that session singers are asked to fill in these situations is often a result of a producer/songwriter lacking experience, who will either consciously or subconsciously co-opt the singer to fill that gap.

> I've noticed that it's more so with people with less experience. They write songs and they do stuff, but they probably have a lot less experience in the industry. And their thing is *I'm going to get this singer in that I really like and she's going to sing on my album.* And they think that it's okay to kind of go, *well in this section I don't know what we're going to do. So, could you just come up with some stuff.*
>
> (AL)

Sometimes you walk into a session where you just get this feeling that wow, these people called me because of my experience over the years to sort of help bring this thing to fruition.

(SW)

Again, it's one of those situations where it's maybe someone is masquerading as a producer songwriter, who is really just a producer. You're not really a songwriter, you know, or if you are, you need to evolve your idea a little more before you call it a song. It's not a song yet, it's half a song, it's an idea for a song. And, yeah, that's frustrating. I mean, it'd be different if you are the intended artist for that song. Then it's up to you to bring it into your own space. That's a kind of a different scenario. But if you're creating this demonstration, then you really want to have a clear-cut idea of what it's going to be.

(JMc)

Crossing the line

While it could be argued that a session singer's unique skillset is *part of the gig* and the reason a singer gets hired in a competitive market in the first place, from the perspective of the vocal booth, there is a clear suggestion that on many occasions a line is crossed. This line is undefined and ambiguous; yet, it exists within a professional commercial setting that uses highly specialised skills and supports livelihoods. Therefore, are these *extra services* at the very least worth acknowledging? Where is this line and can it be defined? For some, the line exists when the level of services rendered is outside and above those that have been either assumed or agreed upon and where remuneration no longer matches the level of these services.

I'm quite happy to cross the line in some cases, but there are times when people have taken advantage of that. But that would be more in the commercial department, in terms of remuneration and acknowledgement. It wasn't there.

(LF)

I do think that part of our role, when we walk into a session is to bring creativity. I think that's just part of the job description. But when you start to write the melody, that's when you have the conversation.

(JM)

Where I find that comes up is you're booked in one capacity as a background vocalist, but then all of a sudden, there's a question of *will you step into this lead vocalist role?* and that is much more of a signature sound, and where it's literally your intellectual property that is being exploited.

(JMc)

It's happened, and it leaves you feeling a little bit used in a sense. You give them 10 hours of work for what should have been a two-hour job.

(GP)

I've talked about this with singers all the time, there's so many lines, and I feel like there's no real governing body to kind of give us advice on what to do when those lines get crossed. I feel like in a lot of other artistic industries, those lines are very, very clear. But as a singer, you just pretty much do whatever you're asked. And I feel like if you do ask too many questions, maybe you won't get booked anymore. If you're going to do an acting job, which I've done many times, there are lines, there are definite lines. If you are artistically asked to do anything else, you can question it, because there are contracts and they're big and thick before you do every job. Whereas, we tend to walk in and just do what we do and walk out. And a lot of the time, there are no rules or regulations on what to follow.

(MS)

I think the line exists for everyone differently, depending on age, experience, how good you are at holding boundaries, how good you are at communicating those boundaries. For example, I was asked to do backing vocals on some guy's album, and what it turned into was me sitting there basically fixing his songs, editing and songwriting. But I actually charged him for it in the end, because I had to go through every single song and fix parts. That's when I had to come in as a co-writer and a producer, and a vocal arranger and all of those things that I had no idea I was going to be doing when I walked in. If what happened to me with this guy had happened even five years ago, I probably wouldn't have had the *kahunas* to come back to him and go, *I'm actually going to charge you for the creative side of this because of the time and because of the nature of what it is*. And yeah, and he was cool with that. And I'm sure there will probably be other people who wouldn't be cool with that.

(NF)

This study clearly demonstrates that there is a lack of system for defining services and remunerating session singers. Fee structures are negotiated on a case-by-case basis, and sometimes, services are renegotiated after the actual recording session has begun. Much of this lack of accountability is explained by the absence of a strong union that also recognises and understands the specific details of session work.

> The union had stopped long before I started, so it was a very different fee structure. People tried to fight for that back in the day. If you did a lead, you got paid. Double, extra fee. Triple, if you were in a section, you know. All that stuff, it was all on a card, and you'd submit that to the accounts, and you'd get paid, like classical musicians. But somewhere along the line that shifted, I don't know when or why, but it did. Certainly in the Sydney scene. The (advertising music) houses created the fees and you either had to say yes or no. Every house had their own sort of story and each artist had their own sort of thing.
>
> (DP)

> It would be really good for people to know, well, this scale works for this to happen, and this is the next scale, and if this has to happen, then it's another scale all together. So that they understand the structure of how the sessions work, and the artistic creative element that comes with it, sometimes by sheer accident, that they've got to be open to. Like a union, right? These are the parameters that we work in as session singers.
>
> (NF)

What these perspectives demonstrate is that while session singers have a personal sense of their professional value within any coproduction event, most lack a model or structure to express their value to the people hiring them. Those that communicate their value, do so *before* the recording session, which enables all parties to start with shared governing variables and avoids confrontation while the session is in progress. This is akin to creating an informal contract that aids in clarifying responsibilities of each of the participants and stands as a shield against co-option and exploitation of extra services.

> In the initial briefing, if a more creative role is required, I always query credit/compensation.
>
> (SA)

I think people got used to that if they rang, I would ask questions. I established that pretty early. You asked the questions before you go, you say, what do you need me to do? *Yeah, it's a session where you're singing on a jingle for a company*. And I go, well that's great. And is it going to be all platforms, like for eternity for the universe? *Oh, yeah, well, we think that's what it is*. Well, you know, define what that is. Get back to me about that. And then, before you start, *will I be doing anything else?*

(DP)

What do you need me to do?

This quote can be summed up in a question that all session singers should ask before stepping into a recording session: *what do you need me to do?* As this chapter demonstrates, the list of possible services may go far beyond just *singing on a jingle for a company*. Using the reflections of participants in this study, I have compiled a list of questions one may ask in relation to a production brief that offers a glimpse into what an informal contract might contain, as well as a list of services that goes some way to answering the question *what do session singers do?* (Table 3.1).

The items in the PRODUCTION BRIEF can be applied to any recording session. The items in EXTRA SERVICES are points that participants recall arising during recording sessions, services that might be considered as co-opting the singer's artistic property. In addition to having informal conversations about the brief before the session, I often advise singers to itemise each of these points on their invoice, which formalises the details of the transaction.

Table 3.1 Roles and responsibilities that manifest in session singing practice

THE PRODUCTION BRIEF
- Who is the client?
- What is the nature of the recording?
- What platforms will it be played on?
- What territories will it be played in?
- What is the term of the publication?
- Lead vocal: single line; multiple tracks (double, triple, quad, etc.); melody, accents and phrasing; voice quality and tone
- Backing vocal: single line; multiple tracks (double, triple, quad, etc.); multiple parts, arrangement, harmonies, accents and phrasing; voice quality and tone

EXTRA SERVICES
- Singing – improvised sections; personally recognisable vocal style and sound.
- Songwriting – intentional contribution to melodic and lyric structure (top-lining).
- Producing – intentional contribution to creative direction and decision making
- Arranging – creation of vocal arrangements and or harmonies

Table 3.2 Services inherent to the session singer's role

EXTRINSIC VOCAL DIRECTION
- Singing – Pitch; rhythm, accents and phrasing; voice quality and tone; position relative to microphone

INTRINSIC VOCAL DIRECTION
- Singing – Fluctuations in pitch; individual flexibility in rhythm, accents and phrasing; voice quality and tone; position relative to microphone
- Songwriting – Subtle variations to melody and lyrics; ad libs
- Production – Decisions based on interpreting the changing governing variables of the session

---

In addition, I have tabled a list of INCLUDED SERVICES (Table 3.2) that participants generally agree are inherent to the role of the session singer in all recording sessions This includes acting on extrinsic vocal direction from the producer, engineer, songwriter or other stakeholders, and also intrinsic vocal direction, which occurs as part of the translation of the extrinsic instructions and includes subtle autonomous decision making, and idiosyncratic variations, mistakes and participatory discrepancies (Keil, 1987, p. 275; Kronengold, 2005).

Summary

Session singers thrive in the creative space of a recording studio. Coproduction affords them an opportunity to practice their highly specialised skills across a range of challenging tasks and environments. While some singers prefer to work *cooperatively*, focusing their time on getting the vocal take as close to what the brief requires and leaving decision-making to those outside the vocal booth, a majority of singers interviewed are enthusiastic about expanding coproduction to a 'shared task' (Rowe, 2020, p. 795) and willingly move in and out of phases of delineated cooperation into a *collaborative* working style.

In general, from the perspective of the vocal booth, relationships are promotive with *competitive* behaviours inhibited (Johnson & Johnson, 2009). Only five participants were able to recall a session experience where the coproduction process was not promotive and nurturing. In cases where there is the *contestation of ideas* (McIntyre, 2008a), participants see this as a promotive environment, but one that needs to be carefully navigated to avoid any negative effects of perceived competition.

With some apparent regularity, singers are also challenged by situations that may cause uncertainty, discomfort and frustration, when they are *co-opted* into the creative space in a measure beyond what they deem is reasonable. There is a line that gets crossed that is difficult to define, but hinges at the point where extra services are either overtly requested while the session is in progress or insinuated and implied over a period of time.

There are two explanations for these unwelcome situations. The first is the level and quality of communication that precedes the recording session, which if refined, may not only help to avoid the line from being crossed but also improve the set of initial goals and governing variables (assumptions, beliefs, and feelings) that each participant needs to share in order to collaborate successfully. The second explanation is reflected in the way expertise is distributed in contemporary coproduction environments, the compromises that are evident when musicians are co-opted into multiple roles, and the potential for diminishing standards of recording production that this type of approach engenders.

Participants identified that these problems often stem from inexperience on the part of the production/songwriting team, who employ experienced session singers to deliver not only the 'extra thing' required to bring a recording to life (Campelo, 2015), but in some cases to complete the work from a songwriting and production perspective. These experiences appear to be tied to producers and songwriters working in smaller home or production studios, which are a result of the *second wave* of recording studio production (see Chapter 1). While this wave has greatly increased the potential for individuals to create music that is of a professional quality without the need for big studios and big teams of people, it has also exposed unskilled workers who are not experienced across the full range of roles and responsibilities and are lacking the technical knowledge and the power to communicate clearly and professionally.

Within any commercial situation, clear communication and proper, informed acknowledgement of expertise and contribution, are required for all participants to uniformly agree on a set amount of remuneration. Whether it be an hourly rate, or a set fee per song, all parties need to agree to what services that rate, or fee represents, and what services it does not. Further, the details of *how, when and where* the recording will be exploited must be transparent to enable an *informed decision* to be made in regard to fee structure. Crossing this line should summon an ethical debate about the way session musicians are remunerated for their roles in the creative coproduction space.

The items in Tables 3.1 and 3.2 provide a starting point for all coproducers in the recording studio to understand the inherent role of the session singer and the extra services they can potentially provide. I argue that if all parties take stock of these points, communication will improve, which in turn, will facilitate more transparent negotiations around remuneration and acknowledgement.

In the coming chapters, I outline how technical environments and the audio engineer's role have changed, and how these changes have impacted coproduction, particularly the extra services session singers are expected to bring to the coproduction process.

NOTE: See Appendix 2 to read an autoethnographic case study that demonstrates how a session singer can be co-opted into the creative space and the effect this can have on trust, regulating emotions, interpreting directions and shaping a singer's approach to the task.

References

Campelo, I. (2015). "That extra thing": The role of session musicians in the recording industry. *Journal on the Art of Record Production, 10* (July 2015). Retrieved from http://arpjournal.com/asarpwp/that-extra-thing-the-role-of-session-musicians-in-the-recording-industry/

Johnson, D. W., & Johnson, R. T. (2009). An educational psychology success story: Social interdependence theory and cooperative learning. *Educational Researcher, 38*(5), 365–379.

Keil, C. (1987). Participatory discrepancies and the power of music. *Cultural Anthropology, 2*(3), 275–283.

Kronengold, C. (2005). Accidents, hooks and theory. *Popular Music, 24*(3), 381–397. doi: 10.1017/s0261143005000589

McIntyre, P. (2008a). The systems model of creativity: Analyzing the distribution of power in the studio. *Journal on the Art of Record Production, 3* (November 2008). Retrieved from www.arpjournal.com/asarpwp/the-systems-model-of-creativity-analyzing-the-distribution-of-power-in-the-studio/

Rowe, N. (2020). The great neoliberal hijack of collaboration: A critical history of group-based learning in tertiary education. *Higher Education Research & Development, 39*(4), 792–805. doi: 10.1080/07294360.2019.1693518

4 Stepping into the sociotechnical space

Recording studios have been described in the literature as 'sociotechnical spaces' (Leyshon, 2009, p. 1315) and 'machinic complexes' (Gibson, 2005, p. 195), which are 'spaces of collaborative relationships, housing assemblages and encounters of bodies and technologies' (Watson & Ward, 2013). In the traditional sense of recording studio production, the audio engineer acts as an interface between the technology and the singer. Their primary responsibility is to set up the technology in order to capture a singer's optimal performance. Engineers control microphone choice and placement, sound inputs and outputs, headphone mixes, and the general technical flow of a recording session.

Audio engineers

Session singers regard this relationship as extremely important, indicating clearly how an engineer's operational and emotive strategies can impact a singer's performance.

> They're the mechanic that gets the engine finely tuned, you know, for the producer to excel. And the vocalist as well. So yeah, very important.
>
> (AD)

> It's a specialist task, being able to listen a lot more to the sound, rather than the overview that the producer was listening for, or the introspective thing that the singer was listening for. Just having that separation of what you're listening to made a huge difference. The hands-on thing is required, but of course, we get their ears as a bonus.
>
> (LF)

> Some engineers are so incredibly quick. And they understand when I say, okay, *save, double* and they just go bang, bang, bang, and it's quick, and you're operating on the same wavelength. Other engineers

DOI: 10.4324/9781003146865-4

are like, *wait a second, I've just got to fade that and I'm gonna cut this* . . . and immediately the session stops. And you have to then reignite that same energy for the next take. And then you have to wait for them to push the button and then you've got to reignite again. And that's painful. But from our perspective, you want to make an engineer feel valued and appreciated, because then they are doing their best work as well. You don't want to have an oppositional relationship with the engineer because that achieves nothing for anybody.

(GP)

Having the right run up. It's so important for preparation, but also just for maintaining that nice rhythm of the recording session. So if you're going to do several takes of the same thing and every time you've got to wait six or eight bars before the drop-in, it can really ruin the flow. And if you're there for three or four hours, that just becomes a really cumbersome use of the time. And it's tiring, as well. So I my favourite kinds of sessions is where it's like, bang, bang, bang. We've got two bars and we don't even have to talk about it, we just know that this is what's going on, and some engineers are great at that.

(JM)

Engineers can glean a lot from a singer in that first half an hour about how they like to work. Because I'm someone who likes to work quickly. I get really frustrated, and I find I can't hold my focus and my ground, when I've got a recording engineer, who is really slow at the stuff that they're doing, or go off on tangents and start talking about stuff while I'm trying to hold five notes in my head.

(NF)

In the studio, time is money and sometimes they'll just say, *go again, go again, go again, go again*, and we're not getting it. Well, maybe you should listen to it, because I've worked in other situations where it's *go, listen, no, got it!*

(JC)

Apart from generating a good workflow, singers appreciate engineers who show concern, empathy and general interest in what the performer needs to be at their best. The emotional labour an engineer performs is fundamental to their effectiveness as a coproducer and highlights the vital role an engineer plays in the organisational system. When this interface is not operating efficiently, it can cause the whole system to not function as it should. The most practical way an engineer can impact the session singer's performance is by providing exactly what they need in the headphone mix.

Headphone mix

Since the advent of sound recording, technological changes have enabled greater sonic fidelity and control over recorded sounds (Théberge, 1997) and musicians have the opportunity to fine-tune the sound they hear in their headphones, that of their own instrument and the musical accompaniment. Expert performers, producers and engineers understand the difference this feedback can make to an outcome. Campelo (2015) points out that a good headphone balance is essential for two of the most important elements in recording – 'punching in and out, as well as doubling tracks'.[1] Thus, the engineer plays an important role in transporting the singer into a *performance space*, where they might feel comfortable and familiar within their social and technical surrounds.

> When you go in, you want to get your headphone level right, the level of your track versus how much of your own vocal you want. I always have one headphone off, you know, to get natural sound, but I don't like to have myself in the headphones because I find that affects pitch. And so I just go off my raw sound and basically have the track in the other side. With an experienced and intuitive engineer, you explain it once, and then you won't have to ask for it again. You'll walk in and it will be perfect.
>
> (GC)

Williams (2009) argues that a good headphone mix can provide a listening experience that encourages creativity. Clare Torry, who sang the solo in Pink Floyd's 'The Great Gig in the Sky' (Wright & Torry, 1973) from their 1973 album *The Dark Side of the Moon*, attributes her highly celebrated performance to the sound in her headphones

> One of my most enduring memories is that there was a lovely can [i.e. headphone] balance. Alan Parsons got a lovely sound on my voice: echoey, but not too echoey. When I closed my eyes – which I always did – it was just all-enveloping; a lovely vocal sound, which for a singer, is always inspirational.
>
> (Harris, 2005)

Thirty years after the recording, Torry claimed a co-writing credit and settled with the band members out of court. All pressings now list Wright/ Torry as the songwriters (Rolling Stone, 2012).

Amidst the sociotechnical restrictions of the vocal booth, singers are required to be *emotive* while surrounded by an artificial performance environment. Their audience is imaginary and its identity, particularly in popular music, is often undefined (Greig, 2009, p. 17). Holding a position on a stationary microphone means that field of vision is somewhat limited to the microphone itself, which could be why so many singers choose to 'perform with their eyes closed, actively listening to the mic-instrument distance instead' (Bates, 2012). Thus, another crucial aspect of the engineer's role is creating a space that inspires the singer to invoke their own unique concept of *audience*.

It's like the audience is on the other side of the headphones. I listen to the track and try to imagine the band.

(JC)

I love working off an audience, that transfer of energy . . . I adore that. If there's no energy, and you're not getting that from the sound engineer or someone who has written the song, you're like, *yeah, that was great!* (sarcasm). Up to a point, the others are like an audience because you want your performance to be really good. And you want them to like it, and they're paying you and you want them to feel good about what you're doing.

(AL)

There is something about singing when another human being is in the room. Like I know that I sing different just knowing that someone's listening. And there's an amazing thing that happens when, on the other side of the glass, there's someone just raising their hands going, *dude, you nailed that, like, that was amazing!*

(AD)

Summary

Despite the clear value an engineer brings to recording studio coproduction, their role has been one of the first to be compromised due to the downsizing of recording studios that has occurred since the *second wave*. With the lone producer/engineer more often filling two roles rather than there being a clear division of labour, specialist skills have given way to cross-over skill sets that previously would not have been tolerated in professional coproduction. That is not to say that these changes have been 100% unsuccessful, as there are many cases where the conflation of multiple roles has worked extremely effectively and enabled production work to continue on reduced budgets.

However, it does beg the question as to whether overall standards of audio engineering, production, performance and songwriting have been compromised as a result, and whether lower standards are now more acceptable. Downsizing that is a result of the *second wave* has led to a proliferation of less traditional structures, and greater ambiguity of roles and responsibilities. The technical changes that have made recording studio production more *convenient* and *cost-effective*, have also led to large-scale cultural change in the domain.

In the next chapter, I describe how the COVID-19 pandemic of 2020 catalysed the *third wave* of recording studio production, further challenging expectations, roles, responsibilities and processes. This was not a technological change but rather a societal transformation that precipitated cultural change.

NOTE: See Appendix 3 to read an autoethnographic case study that demonstrates how a session singer works with a lone producer/engineer to achieve a successful vocal session despite the many sociotechnical compromises they both face.

Note

1 Punching in and out, also referred to as drop-ins, is recording only the parts of the performance that need replacing. Doubling is the act of creating two or more tracks that sound the same and can be played back together to form one performance outcome.

References

Bates, E. (2012). What studios do. *Journal on the Art of Record Production, 7* (November 2012). Retrieved from http://arpjournal.com/asarpwp/what-studios-do/

Campelo, I. (2015). "That extra thing": The role of session musicians in the recording industry. *Journal on the Art of Record Production, 10* (July 2015). Retrieved from http://arpjournal.com/asarpwp/that-extra-thing-the-role-of-session-musicians-in-the-recording-industry/

Gibson, C. (2005). Recording studios: Relational spaces of creativity in the city. *Built Environment, 31*(3), 192–207. Retrieved from www.ingentaconnect.com/content/alex/benv/2005/00000031/00000003/art00003?crawler=true

Greig, D. (2009). Performing for (and against) the microphone. In N. Cook, E. Clarke, D. Leech-Wilkinson, & J. Rink (Eds.), *The Cambridge companion to recorded music* (pp. 16–29). Cambridge: Cambridge University Press.

Harris, J. (2005). *Clare Torry: October 2005 interview* [Blog post]. Retrieved from www.brain-damage.co.uk/other-related-interviews/clare-torry-october-2005-brain-damage-excl-2.html

Leyshon, A. (2009). The software slump?: Digital music, the democratisation of technology, and the decline of the recording studio sector within the musical economy. *Environment and Planning A, 41*(6), 1309–1331. Retrieved from https://journals.sagepub.com/doi/pdf/10.1068/a40352?id=a40352

Rolling Stone. (2012). *Readers' poll: The best vocal performances in rock history.* Retrieved from www.rollingstone.com/music/pictures/readers-poll-the-best-vocal-performances-in-rock-history-20120905/2-pink-floyd-the-great-gig-in-the-sky-0352701

Théberge, P. (1997). *Any sound you can imagine: Making music/consuming technology.* Hanover & London: Wesleyan University Press.

Watson, A., & Ward, J. (2013). Creating the right "vibe": Emotional labour and musical performance in the recording studio. *Environment and Planning A, 45*(12), 2904–2918. Retrieved from https://dspace.lboro.ac.uk/dspace-jspui/bitstream/2134/21404/1/EPA45-208Emotionallabourintherecordingstudio-FINAL.pdf

Williams, A. (2009). "I'm not hearing what you're hearing": The conflict and connection of headphone mixes and multiple audioscapes. *Journal on the Art of Record Production, 4* (October 2009). Retrieved from http://arpjournal.com/asarpwp/"i'm-not-hearing-what-you're-hearing"-the-conflict-and-connection-of-headphone-mixes-and-multiple-audioscapes/

Wright, R., & Torry, C. (1973). The great gig in the sky [Recorded by Pink Floyd]. On The Dark Side of the Moon [LP]. UK: Harvest Records.

5 Stepping into the remote recording space

The movement of the *second wave* away from large purpose-built studios towards smaller home/DIY/production studios in the 1990s afforded singers and musicians the opportunity to set up their own digital home studios and develop basic production and engineering skills. These were considered *demo studios*, suitable for creating musical templates and demonstrations of recordings that would one day be re-recorded by 'real' producers and engineers in 'real' studios when it came time to do the final product. Earlier chapters in this book have highlighted how some demo studios have employed session singers to work on these *pre-productions*. For some time, the 'real' studios retained the traditional triumvirate of producer, engineer and performer, working predominantly in parallel delineated roles. However, as the second wave progressed and budgets declined, large purpose-built studios began to close down, and the more highly equipped smaller home/DIY/production studios became 'real' studios. Audio engineers were no longer in demand, thus eroding the traditional structure of coproduction in the recording studio.

As internet data speeds and online file transfer services dramatically improved during the 2000s, session singers began to record their vocal parts in their own demo studios, transferring completed files to the producer/engineer via the internet. However, projects that required high-quality outcomes would still be 'properly' engineered and produced.

Eventually, during the 2010s, some singers' *demo studios* became well-equipped enough to handle specific tasks such as recording high-quality vocals, becoming in effect, *vocal production suites* to which producers could outsource that part of a recording project. For many session singers, this opened up opportunities to become *vocal producers* and gave them an extra 'string to their bow'. This development further eroded the notion of what the traditional structure of coproduction in the recording studio looks like. Not only had the technologies been democratised to the point where anyone with the means to do so could set up a recording studio, but the long-standing traditional roles of producer and engineer also became democratised and shared amongst the participants, including the session singers.

DOI: 10.4324/9781003146865-5

This boutique arrangement of coproduction became more commonplace in 2020, when the COVID-19 worldwide pandemic caused massive social transformation. Entire cities, states and countries went into lockdown and it was no longer possible for communities to gather in-person. Musicians, producers and engineers were forced to adopt online strategies to coproduce, and a *third wave* of recording studio production began.

As restrictions have eased, the music production industry has, like many other industries, adopted some of the practices that worked during the periods of lockdown and isolation. The main driving forces for these changes are the benefits to producers and clients of *convenience* and *cost-effectiveness*.

> Basically, it's a financial decision, because it's much more economically efficient for them to farm out the work and get the WAVs sent over the internet. Most of the remote work I've done has been for TV, doing all the background vocals and the lead vocals on some things as well. When you're coping with the deadlines of television, there's an economic decision from the TV production company to do it, but it's also much more time efficient for the producer if they just get the file sent in and they just have to mix it.
>
> (SA)

While it is arguable that these benefits are driving a permanent shift to a *third wave* of music production, what is certain is that session singers have confronted the democratisation of roles and responsibilities in a way that has led to wide-spread adoption of new skills, and adaptation of new processes in order to fulfil these *extra responsibilities*. In this study, I asked session singers about their experience of remote recording both prior to COVID-19 and since the pandemic began. In particular, I asked what has been gained and lost in this transition to remote recording.

What has been gained?

Continuity

For some, whose live performance work was interrupted during lockdown and isolation, remote recording afforded them a way to remain engaged with work and now offers new ways to work in the future.

> The benefit is that I've been able to keep working. I have been doing some remote recording, you know, sending off files, a little bit of session stuff and album stuff. And it's saved me during lockdown because it's given me a creative focus. So even though I've missed collaborating and being in the studio with people, it's actually kept

me alive, mentally and in my soul. Because I really love being in my studio and recording at home. And if this is all I did, I'd be so happy.

(JM)

The COVID experience of isolation has meant that this is now very viable, when you're good at what you do. Some singers don't have the recording gear or don't have the expertise to do it, but because I've written and recorded for so very long, it's all part and parcel for me.

(GC)

It's amazing that you can do sessions from all over the world.

(SG)

I mean, look, how wonderful is it, that we can actually create something, what it does is opens you up to opportunity, and to have people on a project that you may never have been able to have before. And that's what's exciting about the new way, it just means that there are less limits to what you can do.

(NF)

Creativity

Remote recording puts control in the hands of the vocalist, and for some session singers, this freedom drives their work.

I think one thing we've gained is a little freedom and creative control. I usually ask people to give me a brief, or they may give me a demo with something on it that they want me to replace. And I'll always ask, *are you open to me having a little bit of creative licence? Or do you want it exactly as you've laid it out here?* So I always try to clarify that because if they're open to me sort of interpreting things a little bit, usually I feel like I can deliver something that is maybe more authentic, because a) I'm doing it sort of my own way, and b) maybe I have a slightly better idea. How many times have I wanted to fix something on a session? It's like, I know it's not great. It's okay. It's there. But it could have been great. So having the opportunity to correct something, if I feel it needs it, I like being able to do that. And yeah, I think in the end, it probably leads to a slightly more polished end product.

(JMc)

You just have more freedom to do what you want to do. And maybe throw in some of your own ideas.

(JD)

As a singer, in the studio I prefer to have my hands on the wheel, because I know how important it is. It's not just singing, it's everything around me now. You know, it's the plugins, it's the right reverb is the right mic. And I don't see myself as a great vocalist, I think I see myself, like, as a whole package.

(AD)

As a singer, I like feeling empowered, that you can record this in a way that you want to, and it can sound how you want it to sound.

(JC)

Convenience

The convenience of remote recording is not only enjoyed by the clients and producers. Not having to travel to a scheduled session was cited as a benefit by many of the session singers.

You consider travel time you're saving yourself a hell of a lot of time in your day. Yeah. And they're saving time where they can be doing other things as well.

(GC)

It's definitely less pressure. You know, I don't have to be at the session at nine o'clock, where I'm still half asleep, and my voice isn't warmed up. I can do it when I'm feeling vibed, when I've got the energy and, you know, you can record when you're at your best.

(JF)

The positives are just convenience. We've gained time, obviously not having to travel, and we know what the environment is.

(SG)

Having created this extra time, some session singers prefer to use this time to work in a way that suits their personal creative style.

I like it because I can take my time, not worrying about wasting somebody else's time and just doing it at your own speed. That's a luxury as well, you know. So I certainly enjoy that.

(AD)

You've got to be the vocalist and be in the moment, and then listen to it with a different set of ears, in a sense. I think always leaving things for a day or two and coming back to it is super important when you're doing your own thing as well. Because when it's fresh in your head, you remember how something feels. And often in a couple of days, you won't necessarily remember how it felt. But you know what you're hearing.

(JC)

In a virtual sense, I definitely prefer to work by myself. It just saves so much time and hassle? Personally, what I like to do is just one take of how I think I would sing something and then *Okay, can you send me notes back on what you think I should change* and whatever. And so I would take that person's notes, and then I would just follow what the person says again, like I would normally in a professional session.

(JD)

But for others, who prefer to work more quickly, one of the greatest benefits is that the entire process can be sped up.

I've been doing a lot more remote recording from my studio and I can smash stuff out so quickly. The advertising side of my session career is all about getting stuff out as quickly as possible. I'm editing, doing all the crossfades and all the other parts, bouncing out vocal stems, to then upload and send back to them. I can do it in half the time it takes to do it with a slower and non-intuitive engineer. Producer, engineer, the singer. You're putting all of those hats on. And yet it's quicker.

(GC)

I don't know exactly what it is that makes it quicker. I think it's, and I'm not like yanking my own chain, but I do know what I'm doing. And without the extraneous stuff of someone else's energy, I can get on with it and make it quicker. I can also concentrate and match things more easily because I'm used to the sound in my own studio. So yeah, I know my voice and I know the way things work, so I can do it more efficiently.

(SA)

While it is important to note all that has been gained in the movement to remote recording, many singers are also extremely clear about what has been lost.

What has been lost?

Interaction

Recording studios can be understood 'as meeting places' (Bates, 2012) and these facilities affect the way people work. Isolation during lockdown prevented opportunities for coproducers to physically interact, and this was felt acutely by session singers who noted the intrinsic value this connection can add to the music being made on the day.

> So what's been lost is the interaction, the camaraderie. You're missing that human connection, where you just look at each other smiling and just love that person that you're working with, and that love translates through the vocal and translates through what they're doing. And so it's missing that magic that really is the essence of music. And I guess that more and more has been drained out of music, if we are doing it from our bedrooms, devoid of the human contact. The human contact is half of the dynamic of music therapy.
>
> (GP)

> Camaraderie and networking, for sure. And visiting other people and having cups of coffee. That's really good. I think it's great to have the experience of singing in different studios, that's awesome. And I do miss that. Especially now in this COVID situation, I don't know how that's gonna happen again.
>
> (SA)

> But what's lost is definitely that connection, which is something that I feed off.
>
> (JC)

> We don't collaborate the same way when we don't feed off each other's energy in a room.
>
> (AD)

> It kind of saddens me in a way. It's technology running away with this, which can be great. But at the same time, we do crave human interaction. It's probably why half of us or more, do what we do. It's an excitement that you don't get if you're not doing it in that way. If there's not then human interaction, it's a very, very different thing. You forsake one thing to gain another, I guess.
>
> (GC)

> I think what we lose is new experiences with different energies, different people, different producers and engineers. Just that life experience of being able to go into different environments, and have this situation happen in your life where these tiny, little minuscule things might happen. And you might think it's just like any other session, but in fact, they're all different. And they all

generally teach you a little bit of something, whether it's negative
or positive. You always walk away feeling something. And I think
that's a big thing that we're missing out on.

(SG)

Networking

As meeting places, recording sessions also provide networking opportunities to meet an array of other singers, musicians, producers, engineers
and executives. These types of encounters not only demonstrate a session
singer's cultural capital, offering the potential for offshoots of future work,
but also help to build individual social capital, which can sometimes be the
difference between getting hired again or not.

If you're in a studio and there are other people there that can see
you can do the job, chances are they'll recommend you for the next
job. There are many times where I've met a new singer and gone,
*oh, this person can really do this, they can remember the parts,
they can sing it back, they can execute it the way it needs to be
sung.* And I'll always recommend them. Whereas I'll probably be
less likely to recommend someone I haven't worked with, because
there's that unknown. And I'm sure the same would be for them
recommending me. But when you've worked with somebody, or
you've seen them, it's networking. Networking is so big.

(MS)

Immediacy

These kinds of social bonds give immediacy to the experience of working with another person, where there are no delays or miscommunications,
and clear decisions are made in the moment. Therefore, some singers see
remote recording as interrupted and disjointed, due to the lack of flow and
in-person collaboration.

On one hand, I think it's amazing. And on the other hand, remote
recording it slow. That's ultimately the thing for me, it's too slow.
You send this off, you get that back and you go, *No, that's not
right, you got to do it again.* You got to wait for someone to actually be ready to do it, then they do it, then because that changes,
maybe the guitar needs to now change. So you go change that. And
then it comes back to me. It feels a bit disjointed.

(NF)

You need the producer there with you because it's a very in the moment process, recording vocals. It's all very *of that moment.* If you have to go back to it after your producer's looked at their Dropbox and made their notes. And then you interpret their notes incorrectly. And it goes on and on and on.

(CS)

Two heads are better than one, I suppose, where someone can give you an outsider perspective on what it sounds like and what it could sound like. There's definitely things other people are going to bring out of you, or give you ideas, even if they say one thing that makes you go, *Oh, actually, maybe I could try it like that.* And it just might make you reflect on a recording you've heard and then you might want to try and mimic that. So I hope that home recording doesn't become the *be all,* because producers have a lot to offer singers for sure.

(JC)

Sometimes with remote recording, we don't get that instant reaction or gratification.

(SW)

There could have been so many differences in the recording, things we could have done differently, we'll never know.

(GP)

You don't have the interaction in the moment to kind of go, *Oh, yeah, that's great. But maybe let's try doing it like this?* Whilst I'm having conversations before I start recording, I'll make sure I get as much information as I need before I hang up and press record, so that I'm doing my best to bridge that gap between the interaction we'd have had if I was in the room with them.

(GC)

Sometimes when I'm doing stuff at home, I get carried away and I add too many things. It just becomes this dense wall of sound or it gets too intense, like too hectic parts. And whereas if I was to work in a studio with someone else, it's like, *Okay, that's enough. Like, you need to relax. That's too much.*

(JD)

Some producers have tried to develop in the moment processes using various face-to-face network communications platforms. For session singers, this practice does not make up for in-person experiences.

When COVID actually hit, it (remote recording) didn't feel so foreign to me. What seemed really foreign was doing actual sessions on Zoom. They would share their screen, and it's the tracking session, and then we'd have a Dropbox. I'd drop my stuff in and, you know, the producer would play around with that.

(JD)

A lot of people just say, *Oh, you've got your own setup at home, fly me over a vocal on this,* or we might do it via zoom, or just over the phone, on FaceTime or something. Even though they're hearing it sort of in the moment as it's done, they're still not in the room. So it's sounds different you know, they don't get the true essence of the way it sounds. And they say, *Okay, just send me the WAV file.* And you send it and they go, *Ah, I didn't hear that when you did it the other day. Ah, could you just change . . . ?* And you go back and forth so many times.

(SW)

I've been doing some sessions on jingles from Sydney. Sometimes I'm like, just leave me to do it and I'll send you back the tracks, but sometimes it's so specific that they have to Zoom me and they watch me do my take and go, *Oh, can you make it more this?* But for me it's like how are you hearing my nuances and things over Zoom? You're not, until I send you back the actual track. I just think sonically the music has changed to suit home studios.

(CS)

Quality and authenticity

Despite the affordances of gaining creative control and extra time to 'get it right', many session singers are still concerned about the quality and authenticity of remote recording.

I've been doing a bunch of remote recordings for people's records and my current setup is off the back of the garage, this sort of add-on extra room, no insulation. The door doesn't really close and there's one seedy little light in the corner. That is where I had my recording setup this year. It's got its own character, I guess. But it's completely non-treated sound wise. I'm used working in incredible studios and now what I'm sending to people is recorded in the garage. I'm just going straight into my Apollo, straight into the computer. So I feel like that seems okay to everyone, everyone's fine with it. And I kind of, I have a problem with that. And I feel like the problem is that most people can't or don't want to tell the difference. Or it's not important.

(JMc)

Sonically things have changed so much in music, if I listen to music that comes out it's so different. It's almost come to a point where the music sounds a certain way, and that sound can be produced at home now.

(CS)

The cons are that now everyone thinks they're a producer and expert. Even just the other day, I featured on a single and I literally recorded it on my iPhone, just singing into the iPhone. And just making sure that it didn't distort and that was about it. It's not the greatest and real sound engineers would listen to it and say it's absolute rubbish. But for the standard of what it is today, I think that's what it is. And I'm not necessarily mad about it. It's just *it is what it is* these days.

(HT)

What a recording studio provides is the professionals involved, an external energy, the building or space itself taking on its own role in the production. The wonderful equipment available, old, warm qualities that are hard to replicate at home especially in this digital age. And just a general loss of connection, without the need for these 'centres' we lose that revolving door for music, artists and creation.

(SG)

I almost feel like I've had an in vitro baby. I know that might sound weird, because it's like it's being made in a test tube. It's like out-side of me in a way, rather than organically created with people around me in the same space.

(NF)

We've lost craftsmanship. I don't think up until digital that the technology had changed the craftsmanship. In losing the engineer, we lose that legacy of the knowledge of all of that stuff.

(DP)

The acknowledgement of losing sonic hi-fidelity and craftsmanship dem-onstrates the connection and awareness some session singers have had with audio engineers and the void that cannot be easily filled by combining this role with other roles – including their own.

I miss the engineer that has just studied engineering. That's why listening to songs in the 80s and 70s, they all sound different, and they're remarkable, you know, you turn them up and you just get lost them. Oh, my goodness, because there's an engineer that has spent hours moving that amp around. It's not a plugin. He hasn't gone click, click *Oh what a great sound*. He spent a couple of days maybe on that vocal sound. I miss that.

(AD)

And I think what's lost is the appreciation of how a proper good engineer and studio are.

(JMc)

A different gig

The losses coproduction has incurred during the *second and third waves* are significant. They have altered the way coproducers work, interact and feel about the nature and quality of that work. While the movement throughout the *second wave* was not perceived as a giant leap into an entirely different form of coproduction, the movement towards remote recording is seen, at least by session singers, as *an entirely different gig*.

They're totally different. I would say they're completely different. I mean, even just from the different set of skills, you know. But it's funny, because it's not as though the producer gets behind the mic and puts down a line and you go with it. If anything, the vocalist is contributing all these extra things, and perhaps there's no credit happening. So maybe it's an expectation that people have just ended up putting on the vocalists. And we're just meant to deal with it.

(SG)

But you know, is this just part of our, our toolkit now? You know, once upon a time, we put the PA in the car, and the mic and the keyboard. And so maybe this is just part of what we do now.

(JM)

Yes and no, like, now it just feels so second nature, like, this is just what I do now. But if I look back two years, I didn't do any of that. I feel like I definitely put more work into sessions now than I ever have before. And yeah, thinking about it now, ah, I actually do a lot. But no it just feels like it's just become part of what I do now.

(JD)

I'm a singer, you know what I mean? But now I need to know how to produce a track, because the session places will be like *Oh, do you have your own studio set up?* Now it's expected of the job, as a part of your job, almost like singing's not enough, not being amazing. You actually need to record yourself and edit the vocals for us and send them through – for nothing. The job description now has changed. And it's evolving. But my bank account is the same.

(CS)

There's been absolutely zero acknowledgment or recognition that there is any kind of difference between what we did before and what we're doing now. In my in my experience, it's just kind of assumed that this is what you do now, everybody should be able to do it. And you get paid the same, or less, because you get to work at home. So what we haven't gained is a heightened level of appreciation for all these extra tools and skills, which are critical to be able to deliver that stuff. I don't get paid a studio fee; I don't get paid an engineering fee. And again, it's not about money, but it's just that there's been no conversation about this. It's just like, now it's expected.

(JMc)

Documenting (more) extra services

Following on from the tables in Chapter 3, which outlined the services session singers provide or may be required to provide, remote recording adds further 'tools and skills' that should be part of any negotiation. Table 5.1 lists the extra information that needs to be detailed in the production brief,

Table 5.1 Extra roles and responsibilities that manifest in remote recording practice

THE PRODUCTION BRIEF (now also includes)
- Engineering – Sample rate and bit depth; vocal chain (i.e., advice on mic, pre-amp, compressor, A/D interface); room reflections (i.e., advice on dampening and absorption).
- Production – Overarching guidelines for take selection and direction of performance

EXTRA SERVICES (now also includes)
- Location – Vocal production studio hire (recording equipment and venue)
- Engineering – setup of vocal chain (i.e., mic, pre-amp, compressor, A/D interface); monitoring of inputs and outputs (including headphone mix)
- Producing – take selection; drop-ins; intentional contribution to creative direction and decision making
- (May also include) Post-production – comping; EQ, tuning and effects; mixing of backing vocal and/or choir stems

interpreted by the session singer and then actioned using the tools and skills that are listed as extra services.

Summary

While the conflation of roles during the *second wave* of recording studio production did not immediately raise alarm bells in regard to the loss of specialist skills and the democratisation of responsibilities, the movement to the *third wave* certainly has. Session singers have been thrust into a new form of work, where singing is not the specialist skill it once was, and expectations are manifesting that they need to now *share the load* when it comes to recording and production.

Up until now, the concept of the traditional coproduction structure remained resilient and steadfast in both the literature, and also in the minds of those who remember working in that way. However, for the next generation of coproducers, this concept will only exist as an ideal, or a version of coproduction that you can imitate if your means allow for that choice.

> I can't even imagine what it's like for a young singer who's never worked in a real studio, but they have only worked in this digital domain virtually or remotely. Unless you're in the top, top, top tier who is still recording in the big studios. You know, that's a very small minority.
>
> (JMc)

Working in 'this digital domain virtually or remotely' will almost certainly be part of what young singers experience as session work early on in their careers. Social lockdowns around the world since 2020 have meant that for most students in high schools and higher education, learning to remote-record has become as important to making music as live performance. So as we stand on this precipice of what *Covid-normal* practice might look like in the future, now is the time to reassess the coproduction processes, the rules of play, and the standards we aspire to.

As we forge a new future, we must heed the messages of what has been lost during the second and third waves of coproduction and should endeavour to retain particular aspects of past practices. In the next chapter, I offer suggestions for how coproduction can evolve without losing too many of the rich and valuable experiences the past provides.

Reference

Bates, E. (2012). What studios do. *Journal on the Art of Record Production, 7* (November 2012). Retrieved from http://arpjournal.com/asarpwp/what-studios-do/

6 What does this mean?

One of the most compelling images from this study is of the session singer, somewhat confused, frustrated and concerned, questioning the integrity, professionalism and fairness of the coproduction process in which they are engaged and asking themselves *What does this mean?*

Acknowledgement

The problems that cause such events appear to stem from a lack of communication and acknowledgement, not only of the specialised skills that a session singer brings to a coproduction, but also the list of additional creative and practical roles and services they *might* provide in a recording session. For most, if not all, session singers interviewed in this study, remuneration is secondary to this issue of acknowledgement. It is difficult, if not impossible, to fairly negotiate a fee when the goods and services have not yet been clearly articulated.

Session singers have a tacit understanding of their value, but even amongst this cohort of professionals, many have been surprised by their own vague notions of what they do. By itemising these services, session singers can begin to evidence that value, firstly to themselves and then to others, and have that value rightfully and respectfully acknowledged.

That said, it is unclear even amongst session singers how such acknowledgement should be tabled.

> I think it would be quite hard. Unless it was very clear, yeah, to sort of label the contribution that I made, because sometimes I do feel like singers bring a lot of subtle things, so much actually, there's so much I think going on.
>
> (SG)

> I always felt that if you hired the right people, you didn't have to do much. And I guess, the art form of being a good producer

DOI: 10.4324/9781003146865-6

is being able to lay out the parameters and have the right people there that can do the task. And acknowledgement of those skills is really important. Where do you publish those sorts of acknowledgements?

(LF)

Publications such as this book will hopefully begin the process of tabling the work of session singers, albeit in a more general sense. Through evidenced discussion, more stakeholders will be made aware of the issues, and it won't be left to just a few to change the way coproduction functions in the future. Structures for fair and ethical practices, acknowledgement of services and remuneration can be achieved if industry bodies in Australia such as APRA/AMCOS, PPCA and MEAA join this conversation. In the immediate term, for the individual worker, services and contributions can be documented in pre-production discussions and emails and then be confirmed in writing on the invoice. This is the first step to raising awareness amongst those working in the industry.

The next step is to encourage participants to improve their skill sets in a quickly evolving field, and for all coproducers to explicitly acknowledge the time, effort and financial cost involved in developing a professional standard of service.

I would definitely not be charging the same things that I would if I was going into the studio as a singer, hands behind my back, *what do you want me to sing?* That's a different price. And it should be. For people that are doing what I'm doing, sending off files and even bouncing stuff out, you know, and adding effects reverbs. That's a massive thing that not everybody knows how to do properly. So that should be recognised. There is nothing wrong with being a singer that doesn't do that. But someone that is a singer that does the extra work and, you know, engineers themselves should be recognised because not everyone can do it right. It takes money and experience.

(AD)

If there was even an aspirational thing for people to have in that regard. So in other words, if I upskill, if I invest in my studio, if I learn how to do everything really, really, really well, and even push that ceiling a bit, even break through it, you know, then I'll be rewarded. Whether that be getting accolades, or whether it be through money, that might be an aspirational mechanism for people to actually push through that barrier and create something new in that middle space again. Even remote.

(JMc)

Negotiation

Documented acknowledgement not only provides evidence of the 'services rendered', but also supports and incentivises the negotiation process. In such an adaptive environment, participants have been programmed to *go with the flow*, which has led to many of the issues outlined in this book. For too long, there has been little or no systematic negotiation in this field of industry, and little or no willingness to point that out. However, you cannot negotiate unless you know precisely the points you are negotiating. These perspectives from the vocal booth have offered a table of items (Tables 3.1 and 5.1) that can facilitate open and transparent communication, supply acknowledgement of the practical services and creative input provided and inform the negotiation of remuneration.

As a university lecturer, I have also become accustomed to receiving emails such as the one below, from students starting out in their session music careers. I use this example to demonstrate how acknowledgement and communication are central to negotiation and getting off to a good start.

Hi Rod!! I had something I wanted to run by you/get some advice on if that was okay (I feel you'd definitely be the person who knows most about this kind of thing)!! I have been approached to do some regular recording from home for a royalty free music library. I won't be required to write anything as it will all be pre-written and I think most tracks won't be a full song's length. I won't be getting credited and will receive a one-off payment per track of $150. From my understanding the tracks will be used in commercial soundtracks. Is there anything here that sticks out to you as possibly problematic or something I should address? Or are there any particular questions I should be asking before I start working for them? Basically I will be working through a manager who has an account with this library, so it won't be 'my' music or under my name.

My reply.

Hi!! It is great that you are able to define some parts of the brief, but there are still many more questions that should be answered before the fee gets negotiated. I would start by saying to the production house, 'I would just like to be totally clear about what it is that you want me to do'. Then you can use the following list of questions to cover most of the bases.

Is it a single line lead vocal or multiple tracks (double, triple, quad etc.)?

Are there backing vocal parts, and will these be multi-tracked (double etc.)?

Will you provide me with any backing vocal arrangements? Or do you want me to write these arrangements for you?

Will I be contributing to any songwriting? (this looks like it's been covered, but you might as well double check)

Do you need me to imitate a particular vocal sound?

Are you happy to use the existing setup in my recording studio?

Do you want me to engineer and produce the final vocal remotely?

Is there any post-production you require me to do (e.g., comping; EQ, tuning and effects; mixing of backing vocal and/or choir stems)?

May I add this job to my CV?

It is possible they'll say *look, I can't be sure because each track will be different.* But that will be an opportunity to say, 'Well I'm happy to commit to x, y and z for this agreed fee'. Then you will have some room to negotiate if anything changes. Make sure to list each item separately on your invoice.

I am not suggesting here that there is a specific formula for negotiating a fee, but rather a structure for having a reasonable conversation about the participant's contribution to a project that will not only *clear the air* and avoid confrontation but will also facilitate better communication and more effective coproduction.

The challenge in negotiation is for all participants to accept that in some scenarios, the traditional roles and responsibilities have been so widely democratised that it represents a different form of recording studio coproduction and requires a different way of thinking about processes, acknowledgement and remuneration. It is no longer safe to assume the roles and responsibilities a 'hired gun' will provide. All roles need to be interrogated, *will I be producing, will I be engineering, will I be songwriting, will I be arranging, will I be hiring my studio?*

The future

Coproduction in the recording studio has evolved significantly in the past 25 years, and the movement to remote recording has further heightened delineated cooperation and the democratisation of roles and responsibilities. I predict that remote recording will continue to advance and improve and that it will be part of a hybrid form of coproduction where participants combine in-person collaboration in centralised, larger production studios with parallel autonomous remote work. Across the industry, participants will need to develop their *cultural capital* in the form of cross-disciplinary skillsets, aspirational notions of quality and authenticity, and ethical structures that support the industry in terms of acknowledgement and remuneration for services.

Social aspects of recording studio coproduction such as interaction and networking must also be nurtured. The trend towards isolated parallel cooperation in remote satellite structures has certain benefits, but there is much to be gained by pursuing relationships in the field. Finding new ways to create *meeting places* in which to interact with peers and colleagues with a view to learning and understanding the craft, and allowing oneself to be guided and mentored by those who have a connection to past practices would provide invaluable insights and networking opportunities, and promote cultural growth in the sector.

The sociotechnical challenges that still lay ahead are numerous. Can we develop person-to-person over IP technologies that enable real-time remote collaboration? Can we use the technology to enhance creativity rather than homogenise the outcomes? Will the specialised role of the engineer return in some other form? Will we see COVID-19 as an opportunity to reset this part of the industry and in the fightback to covid-normal, restore some of the qualities that have been lost from the past?

Conclusion

The practice of session singing is experienced differently in every recording session as a result of a number of variable aspects: the relative capital possessed by each of the participants and the associated power dynamics; the emotional empathy and trust that is performed by each of the participants during the session; the communication of the brief and assignment of roles and responsibilities; the willingness and ability to adapt when the functions of clarity, communication, capacity, empathy, technology, artistic property, roles and responsibilities unexpectedly change; and the acknowledgement of these changes and what that means for the session singer.

The session singer's need to adapt evidences the high degree of fluidity that is present in recording studio production. But it also highlights that one of the main causes of this fluidity is the often-prevalent lack of transparency and accountability between the participants. Collaborative dexterity (Rowe, 2020) is an extremely valuable skill in recording studio coproduction, but

I believe it can be much more effective and focused within a clear, fair and rigorous structure based on communication and acknowledgement of services. Such a framework should stem from ongoing discussion and debate about the way we in the industry want to see coproduction conducted in the future.

Coproduction in the recording studio can teach us a great deal about organisational practices. Applying many different examples of collaboration, competition, cooperation and co-option that occur in the structured organisational environment of the recording studio can highlight the skills needed to coproduce effectively in any field. Teaching recording studio coproduction in a tertiary educational institution has not only offered me an opportunity to inform future music professionals, but to teach collaboration in structured and situated context.

Finally, these questions don't just apply to session singers but all session musicians who are also impacted by the issues laid out in this book. It is my hope that further research will develop more nuanced answers to all of the unanswered questions as more perspectives are studied.

Appendix 1
Case study 1: empathy

In 2006, I performed a vocal recording session for renowned Australian songwriter/producer Graeham Goble, recording choir parts on his solo album project Let It Rain (2008). The event was memorable, as it was the only time in my career as a professional vocalist that I left a session feeling as if I had not adequately completed the task. Graeham has an unusual methodology for recording vocals, and at the time I found it both difficult and exhausting. I was unable to get the result he was looking for, which confused me, because at the time I believed I could adapt to any vocal booth situation. Now, 12 years later, I have the opportunity to work with Graeham again, to apply what I have learnt in the intervening years and, perhaps, to understand something of that session – and what went wrong. I didn't know whether it would be a similar experience or not, but I was willing to approach it with an open and inquisitive mind. Graeham needed some assurances as well, inviting me to a meeting where we could 'test out the voice' and see if it was suitable for the task. I had no issues with this test, as it presented a mutual and transparent opportunity to clearly understand the project and all it entailed.

We met at his house in a space that functions as a recording studio control room (figure A1.1). Gold and platinum albums adorn the walls of Graeham Goble's home studio, acknowledging his success as a songwriter, recording artist and producer. Prominent is his work as a founding member of the world-famous Australian rock band Little River Band, hereafter referred to as LRB. It is an intimidating sight and a reminder that I am working with someone who has extremely high achievements and even higher expectations.

Graeham explains that today's project is a theatrical adaptation of the songs of LRB, a cast album, demonstrating the songs in their theatrical context. Graeham wants me to record *The Night Owls* (Goble, 1981), which was originally sung by the band's bass player Wayne Nelson and produced by legendary Beatles producer, Sir George Martin.

Figure A1.1 Graeham Goble sitting at his recording console.

Source: Photo courtesy of Graeham Goble.

Behind the recording console sits Graeham's son Nathan, an accomplished musician, who is today, playing the role of the engineer. Graeham gratefully employs his sons, Josh and Nathan, to engineer his projects:

> 'I could do it [engineer], but I've got to listen and concentrate on what you're doing . . . It's not just the engineering, Nate and Josh are such incredible musicians, they come up with inspired ideas. They're more than pushing buttons and engineering'.
>
> 'We're going to record it line-by-line. *There's a bar right across the street*, we'll do it, then move onto the next line, and the next line'. Graeham explains that he is using this approach because he has got to get me 'as close to Wayne's performance on *Night Owls* . . . I want to have a cast album that sounds as good as the LRB recordings'.

Another feature of Graeham's recording methodology involves creating a recording loop in the software program that enables me to repeat the line multiple times – as many as 15 times in the one take. These recordings provide Graeham with an array of options from which he can select the best performances:

I might be looking for a tail . . . suddenly you do the right tail. . . . When you're doing a line, I am saying 'he's not doing the articulation, the pronunciation on the word I'm, I haven't got that yet', and then suddenly you'll do one. . . . It's about me scanning that line, this is why I'm pretty exhausted when we're finished because I'm listening to 'how's the pickup? Is it clean? Is it in tune enough? Is it trailing perfectly in tune?

I ask, 'so you construct a full performance out of multiple takes on every level? Phrases, words, onsets, tails.'

Far better than you would ever get on one performance. There is no such thing as absolute perfection. And a lot of people misunderstand me, saying 'you're a perfectionist, you're too hard to work with'. It's not about that. It's only about getting it as good as we can get . . . given the singer, the day, the material he's got to work with, all these parameters, this is as good as we can get.

While there is no mistaking that Graeham controls the power in this recording session, he is also sympathetic to the vocal booth performer and the conditions they are working in. But then he also demands their best and nothing less.

Passage from the control room to the recording booth is down two flights of stairs into a basement space where the booth stands in a corner. Inside the booth, a lamp illuminates the music stand, while the remainder of the space is visibly in shadow and audibly 'dead'. The acoustic characteristics are a familiar aspect of traditional vocal booth designs, but the lack of light (a by-product of the design) takes some time to get used to. While not as aesthetically appealing as the upstairs space, the booth is functional, providing a suitable acoustic environment for the AKG C12 microphone that Graeham explains had been used in the 1980s to record Barbra Streisand (figure A1.2).

'OK, let's get a sound', says Nathan. He plays the first verse and, as I begin to sing, I am instantly aware of the power of this microphone. Its clarity, regardless of where I stand in the booth, picks up every sound of my voice, lips, tongue and breath. Graeham gives me a firm instruction:

Once you get your position, keep reminding yourself not go off the position. I always mention that by habit because, I know you are very experienced, but some singers just don't concentrate on where they need to be, and an inch makes a difference with the sound of the voice.

Figure A1.2 Studio of Graeham Goble. The centrepiece of the vocal booth – AKG
  C12 microphone.

Without the usual sight lines between the vocal booth and control room, I feel even more isolated. When the talkback button is pressed, I overhear their conversation in my headphones.

NATHAN: Sounds good
GRAEHAM: And the level of energy, projection and power? What do you think of the tone?
NATHAN: Yep, sounds good.
GRAEHAM: OK, then. Let's go.

After completing the first verse line-by-line, we moved straight on to verse 2. The following is an account of recording one line from towards the end of this verse.

Verse 2: lines 21 and 22 – to rock 'n roll, they always will

'OK line 21. We'll just do the first half, *They have a need*'. Line-by-line recording is demanding on the body and mind, but now Graeham wants to split lines into two. My initial feeling is that this is going too far, and I am unhappy about it. With no visual sight line to the control room, I express this physically, throwing my hands in the air and pulling a face, something I would never do if the producer could see me. Still, I put my emotions in check and resolve to completing the task as efficiently as I can. We do one burst of ten takes and the line is done.

I am pleased to hear that Graeham wants to do the second half of the phrase in conjunction with the following line. Breaking up a phrase can sometimes work, but often it feels unnatural. While Graeham is consistent in his method, he is also sensitive to such details in the singer's performance. Without this *technical empathy*, I would have to either explain myself to the producer or refrain from commenting at all.

'The start of this line is "in the pocket" but the end is more broad"'. I have heard these terms used throughout the session and understand them to mean 'in sync with the rhythm' and 'more relaxed within the rhythm', respectively. I have heard Graeham say that 'a lot of people who sing this song don't get that shuffle feel in the vocal . . . Wayne sings it like a bass player'. While it has been some time since I recorded a song with a shuffle feel, my experience enables me to understand and agree with what the producer is saying. 'Keep the energy up. We need that presence and power right through the line'.

Energy, presence, power. I can create energy by varying where I place each note in relation to the beat. On, or slightly ahead of the beat can give a sense of 'drive'. Sing behind the beat and the song can sound like it is slowing down. Vary the placement slightly throughout the phrase and it can sound more 'relaxed'. Presence and power are not about volume or loudness. It is about concentrating on the line from start to finish, maintaining posture, placement at the microphone, vocal technique and a suitable articulation and vowel shape. I know that if I lose concentration and one of these elements is not there, my presence and power in the recording will diminish. Listening to the line, I decide that *to rock 'n roll* sounds better when I articulate and accent the words right on the beat, while *they always will* can be more connected, less articulated and more relaxed in terms of timing.

On the first burst I complete five takes, aiming to keep the line clean and strong, backing off the air pressure, keeping my vocal tract wide open and maintaining my position on the microphone. Take 1 starts on an A3, leaping a minor 7th to G4. On every other take, the interval is an octave plus a minor 3rd, starting on the E3. While all of my focus has been on the high parts, I haven't paid attention to the entry note.

'Can we listen to the pickup *to*? You're sort of throwing it away and I think you can give it just a little bit more. I understand you've gotta get up to the high note'. Once again Graeham is thinking like a singer. Without his empathy, this method would be much more difficult to administer. Nathan adds, 'Wayne also sang "*ta*" instead of "*to*"'.

While Nathan's input into the conversation is minimal, his comments are always useful. It is a quality that makes an engineer more valuable to the vocal booth performer than one who speaks regularly but with comments that only cloud the minds of the producer and performer. It creates a feeling of order and teamwork in the recording studio.

While the major section of the phrase was consistent over the eight takes, differences were noticeable at the beginning and ends of each line.

The entry note varied between the lower E3 and a G3, creating an initial interval of an octave. The cut-offs varied as well, with each tail varying in intensity as a result. Between each take I can hear Graeham and Nathan's comments, which are aimed at guiding my performance.

'We've probably got this, but I just want to try for a bit more "angst" on the word *roll*. I can't do it, but I can hear that there could be a bit more "street-ness" in the word *roll*. I think we've probably got the line anyway, but we might be able to get something even better.'

While Graeham cannot demonstrate, I am confident that he is looking for some 'growl', a vocal style element common in rock music.

On take 1, I add a slight amount of growl and vibrato to the *roll* and also the high note of the syllable *al-*. 'That's the sort of idea', says Graeham, encouraged but sounding unconvinced. On take 2, I add growl to the word *rock*, and accent the high note more prominently. My physical gestures become more pronounced as I conduct the pitch with my hands. 'Probably a bit rough now'. I now know the parameters that I am working in and over the next three takes I slowly back off the intensity of the growl until I hear 'Two more. That was good, that one'. After two more takes, I hear 'one more . . . Ok. I reckon we got that'.

Discussion

The distribution of power in this session was clearly weighted towards Graeham, whose cultural capital within the domain of recording studio production and vocal booth performance is built upon years of experience in one of the most successful vocal bands in music history, working with the very best producers, engineers and musicians with the best technologies available at the time. In this regard, Graeham possesses a degree of symbolic capital that reinforces his 'ability to wield power' (McIntyre, 2008a). Each participants' role was clearly defined: Graeham (producer), Nathan (engineer) and me (performer). The presence of a dedicated engineer created a more traditional dynamic, in which the engineer and producer could discuss performances and then direct their summary findings to the performer. While this effectively disempowers the vocal booth performer in one respect, in another it enables them to focus solely on singing.

The vocal booth comprised a number of aspects of material productivity (Bates, 2012; Gibson, 2005; Leyshon, 2009). It was a dark, confined, isolated and unobservable container – a tomb, into which a voice from 'above and beyond' dictated my movements. The placement of the booth in the basement of the building can also be figuratively related to the distribution of power, separated by elevation and a relatively long walk, which meant that once I was there, I was there for some time. With no line-of-sight, we relied only on paths of audition for the transmission of information, and any breakdown in theses paths left me isolated. Being unobserved, I was free to act and gesticulate in any way I wished, with no feeling of self-consciousness. But this also affected the way I regulated my emotions throughout the long session.

Although Graeham had given strict instructions about holding my position on the microphone, the 'dead' acoustics and the quality of the equipment actually made this less of a problem than in other more 'live' spaces. Concentrating only on my material surrounds created a claustrophobic feeling but focusing instead on the sound of the C12 microphone provided a wonderful sonic space in which to perform.

Graeham employed a mixture of emotional empathy and neutrality throughout the session (Watson & Ward, 2013). Without the empathy and expert understanding of vocal booth performance, the demands of his chosen methodology could leave a singer bereft of the emotion needed for the performance. Despite his microscopic attention to detail, Graeham's main concern was emotion or 'magic'. My response towards his instructions was generally emotional neutrality, which aimed to keep any negativity in check and save my emotional expression for each performance.

The methodology of short bursts of multiple takes places high demands upon memory, concentration, reflective thinking (between takes) and emotional and physical regulation. It was significant that Graeham acknowledged these demands from the outset as it established the transmission of empathy and trust between us. Without clear and unambiguous transfer of information, I would be isolated and disorientated, physically, technically and emotionally. I argue that these are make-or-break moments in vocal booth performance, when one's capacity for emotional regulation and trust in the other participants either derails a session or gets it quickly back on track. I also contend that this was the main difference between the current session and the one 12 years ago.

Within each recording session scenario, power distributions affect the way participants communicate and negotiate. They can also affect the degree to which participants are willing to accommodate and/or adapt their own particular modus operandi to that of others or the creative system as a whole. Twelve years earlier, I possessed less cultural and social capital than that which I had accumulated in the intervening years. This affected the earlier session in two ways: firstly, I was in an even less powerful situation than that in current study, and secondly, I was not equipped with the emotional and physical regulatory skills that such a session required.

Investigation of the way recording sessions work enables participants in other creative systems to better understand the operation and function of the environment and each individual therein. In a studio world in which the producer, engineer and performer assume clear roles, and in which the relative power dynamics are clearly apparent, and where communication and empathy are clearly transmitted, the vocal booth performer can be fully applied to the task of delivering a vocal that satisfies the client. Memory, concentration, reflection, technical, emotional and physical management become more focused, enabling a performance that reaches towards the optimal balance of technique and art.

Epilogue

The written account of this session was sent to Graeham as part of the ethical research process. About 18 months later, I received an invitation to see Graeham's new vocal booth, which he built in response to my recorded observations. He thanked me for the feedback and the results speak for themselves (Figure A1.3). Apart from increasing the size of space and

Figure A1.3 Studio of Graeham Goble. The newly renovated vocal booth that Graeham built as a direct result of the feedback from this study.

Source: Photo courtesy of Graeham Goble.

providing more options for lighting in the room, the booth now has visual sight lines to the control room via CCTV cameras and a more comfortable aesthetic appearance.

References

Bates, E. (2012). What studios do. *Journal on the Art of Record Production, 7* (November 2012). Retrieved from http://arpjournal.com/asarpwp/what-studios-do/

Gibson, C. (2005). Recording studios: Relational spaces of creativity in the city. *Built Environment, 31*(3), 192–207. Retrieved from https://www.ingentaconnect. com/content/alex/benv/2005/00000031/00000003/art00003?crawler=true

Goble, G. (1981). The Night Owls: On Time Exposure [LP]. Australia: Capitol Records.

Goble, G. (2008). Let It Rain [CD]. Australia: Words & Music.

Leyshon, A. (2009). The software slump?: Digital music, the democratisation of technology, and the decline of the recording studio sector within the musical economy. *Environment and Planning A, 41*(6), 1309–1331. Retrieved from https://journals.sagepub.com/doi/pdf/10.1068/a40352?id=a40352

McIntyre, P. (2008a). The systems model of creativity: Analyzing the distribution of power in the studio. *Journal on the Art of Record Production, 3* (November 2008). Retrieved from www.arpjournal.com/asarpwp/the-systems-model-of-creativity-analyzing-the-distribution-of-power-in-the-studio/

Watson, A., & Ward, J. (2013). Creating the right "vibe": Emotional labour and musical performance in the recording studio. *Environment and Planning A, 45*(12), 2904–2918. Retrieved from https://dspace.lboro.ac.uk/dspace-jspui/bit stream/2134/21404/1/EPA45-208Emotionallabourintherecordingstudio-FINAL.pdf

Appendix 2
Case study 2: co-option

Summer in Melbourne. It's early January, a time when many Australians are away on holidays and most musicians are out of work. It's a time when my body feels refreshed and craves the sun and exercise, and I'm afforded the luxury of time for both. There's no rush to get to the recording session today, no competition for a car park and time enough for a visit to the coffee shop across the road from the studio. The Base is a purpose-built recording space in a district of refurbished factories. It has three main rooms. In the entrance space, half a dozen people (probably musicians, engineers, composers, or arrangers) are working away on their laptops, listening on headphones and oblivious to everything else around them. Through the back of this room is the main recording area, a traditional-style control room that looks out onto an acoustically isolated rectangular recording space through a large double-pane glass window. It's a design that I'm very familiar with, reflecting the style of many larger recording studios that once operated in this district. The Base is small in comparison, but perfect for a solo vocal recording session.

The venue was selected by the session engineer, Jared Haschek. On any other occasion, he might be one of those people working away in the entrance space, arranging music for a client. It is familiar to me as well. I worked here a few years ago and remember it as a comfortable and practical place for recording vocals. The session has been coordinated by the producer Jason Simmonds, director of a large amateur community choir in the north-east of Melbourne, the Melbourne Contemporary Choir. Jason has asked me to record a song called 'The Others' for an EP that the choir will release commercially. He tells me I was chosen because of my connection to the song, an ode to the plight of the socially downtrodden. Jason grew up being involved in a socially conscious organization known as the Salvation Army. So did I, and there's the connection.

'You've done a lot more recording than I have, so I'm really interested in hearing your creative spin on things.'

Jason also knows my professional experience as a recording vocalist. I sense that he not only trusts, but also hopes that I will be active in the creative choices made during the session. Jared is an accomplished musician and arranger and is officially the project's musical director. Although he sits in the main chair behind the recording console, he admits that his role on the day is more musical than technical, mentioning his 'limitations' as an engineer on three separate occasions.

> I'm not a mixing engineer, so I'll just record it dry and flat . . . I'll do what I can but because of my limitations, you know . . . if the person on the mic is quite particular about the sound of the vocal, the solution might be beyond my abilities, I guess to be honest.

He will produce the session and press record, but many of the technical decisions such as microphone type, placement, the routing of the audio signal and the setup of the headphone mix have been made for him by the venue's in-house manager.

Jared's arranging skills are on display when he presents me with a clear lead sheet and piano chart of the song.

'Do you want to talk through the sections, or . . . ? Let's do that', he exclaims enthusiastically.

Jared explains that each of the sections – verse, pre-chorus, chorus, and bridge – were originally written with choir and both male and female soloists in mind, and the first verse is in an alto range.

'The arrangement is what it is. I would have done things differently if I was living my life again . . . it's all good'"

I am concerned that each of the sections has very clear endpoints, they are very distinct in their vocal range and approach and we have to tie the meaning of these changes together, so it makes sense performance-wise. In the studio environment where a performance will be preserved, and become, in effect, the definitive work, these factors play an important role.

'What if we let Rod do the verse and pre-chorus and then we let the choir take the first chorus?'

Jared is starting to get organised.

'For this first part, I don't care what you do with it. If you want to change it rhythmically or melodically, then . . . all ears.'

My preferred approach is a little more conservative. 'What I'll do is, I'll start with the structures, as written, and then branch out a little bit. You can always stop and start me again'.

The responsibility of taking an arrangement written for multiple voice types and creating a lead vocal part that is threaded seamlessly throughout the song is, for the moment, the biggest challenge of this session. Although

I am afforded the freedom to explore and invent new melodies, the arrangement is also the greatest limiting factor thus far. While it is well structured, the adaptation for a lead vocalist is not fully formed. Within this situation, I feel some pressure, but I am empowered with agency to voice my creative and technical opinions.

Jason and Jared trust me, and that is a position in which I feel very comfortable. However, their admissions to being limited in the roles of producer and engineer, respectively, make me think that the inevitable problem-solving that occurs in a recording session could fall back onto myself or perhaps others if things get too complicated. My trust in them is not as great as their trust in me. Still, we're getting along and that's the most important aspect at this point.

As I move from the control room into the tracking studio, I begin to feel more comfortable. One of my favourite microphones, a Neumann U87, is positioned in the middle of a large room, alongside a music stand and a Behringer Powerplay 16 headphone amplifier that adjusts the overall volume of the AKG K77 headphones (Figure A2.1). I prefer to be able to control the mix of the backing track and my own vocal, and it occurs to me that, most of the time, engineers create these workspaces without enquiring about a singer's preferred setup.

Figure A2.1 The Base. The view from the singing position into the control room.

By the time the session begins, and the clock starts ticking, it's too late for any last-minute changes. I can see straight into the control room through a large window and there is also a grand piano in the corner of the room. I am familiar and comfortable with this setup and this alleviates my tensions. I also notice that the coffee I drank prior to arrival is activating in my stomach, which is not a good thing when you are about to sing.

'Do we want to aim to get an edited-together fantastic take or do we want to get, like, you know, a few great takes and put together . . . ?'

I assume from this comment that the vocal takes we record today will be 'comped', in which different takes are cut and pasted to form a final recorded work.

'Shall we give it a go? Let's record up to the end of the first chorus and see how it shapes up'.

Discussion

This account of the pre-recording phase highlights several explicit aspects of recording studio practice that inform baseline governing variables and, in turn, the initial recorded performance – Take 1. Jason and Jared espouse that they are happy to defer to my way of working in the studio, that they are operating in roles somewhat unfamiliar to them, and that there are limitations to their skills that might affect how the session progresses. This information establishes my belief that the relative power balance in this system has shifted towards me and that the roles of engineer and producer will be shared by all participants, including myself. These types of non-standard sociocultural recording studio parameters affect vocal booth performance by placing additional roles and responsibilities on the performer that must be enacted in parallel with their primary task of singing the song.

The requirement to adapt a choir arrangement to a lead vocal performance while in the process of recording adds a further complication to the task by placing the vocal booth performer in the additional role of arranger. There is no specific reference to the field of works, and I assume that this song is in the general pop/contemporary/gospel category of music styles with which I am familiar.

The discussion concerning 'my way of working in the studio' establishes that I can direct the initial recording methodology and objectives. The one instruction from Jason to 'exhibit a connection between the vocal and the lyric' places emphasis on my belief that the greatest challenge of this session will be to maintain or develop the thread of this vocal performance throughout the sectional transitions. This means that we will work in larger cross-sectional chunks, with the aim to create seamless changes between the contrasting sections. Finally, the intention to 'comp' the vocal later establishes the belief that we are not expecting to get a single perfect take of a song, section or even a phrase. The process is to select and collect sub-sections of each take that will contribute to a final performance outcome. In this situation, the vocal booth performer believes they will be able to provide takes that are singularly selectable, with the knowledge that the decision is ultimately out of their hands.

Recording phase: the bridge

There were nine separate takes of the bridge section of 'The Others'.

'I'm marking the good takes as we go', says Jared.

This is a practice that technicians use to make their comping a lot easier later on. It can help to keep track of what has been selected, so the focus can go onto the sections that still need work.

'The sopranos take the bridge. Sorry, I didn't write this for a guy'.

Looking at the score, I can see that the bridge is written in a very high range.

'I've got to sing this right up there? How do you want me to sing this?'

'That was going to be my question . . . ' Jared responds.

'I'll let you sing it and then we can talk about it, might be a better way'.

'We've just got to remember to keep the choir in mind,' Jason adds.

The actions of others inform my approach. If the producer is choosing to identify usable takes rather than one complete take, I have the freedom to explore variations on each repetition, offering up different ideas within the section that the producer may select. The onus is therefore on the producer to communicate to the performer if their methodology changes and a more imitative approach is required. I sense that Jared wants to offer clear direction, but he has decided to default to my interpretation and problem-solving techniques, which begin with singing what is written and then varying the outcome until everyone agrees that it is satisfactory. I also have to remember the choir and how their phrasing might sound in relation to my part. We're now an hour and a half into the session, and after an apprehensive start, a veil of bravado is descending on me. I'm warming up both vocally and to the challenge of showing off my range, and I approach this sight-reading exercise with the intention of a fully expressive outcome, negotiating my vocal technique and demonstrating what a session singer can achieve in one take.

I do my best with the first take, but the written melody reaches the upper-most part of my register and I struggle to sing the notes clearly and with control. An error exacerbates this. I knew that it was going to be difficult, but I wanted to try, feeling as though I had something to prove to myself and to the other participants.

The result confirms to everyone that we need a dialogue and a change in strategy for this section. There are no suggestions that I should find an alternative melody, but the general assumption during this session has been that we will 'know' when we hear it, rather than imagining it in theory. That means that I need to bring the ideas to be critiqued.

After the poor outcome of take 1, I hear some doubt in Jared's comment.

'Is this the point where we let the choir take over?'

I assert my own opinion.

'We're talking about trying to connect all of these jigsaw pieces. What about if I construct my own journey through this bridge?'

He hastily agrees.

I believe I have been given freedom here to do whatever I wish, to exhaust all possibilities and deliver a vocal that my clients would not get from another performer. I have been let off the leash and I subtly feel my ego taking hold. I study the chords written on the score, and a mixture of excitement and worry creeps over me. Again, I know I have the skills to quickly arrange a melody for the lead vocal, but am I wasting everyone's time? Time is money in a commercial studio. I am now cast in the role of arranger and improviser, limited in a sense by the inaudible choir who will be recorded later.

On the second take, I improvise a melody while looking at the sheet music and following some of the facets of key lines. In my body, I can anticipate when the high notes are coming; it's as if my body prepares for them, but I choose to sing a lower harmony, perhaps in contrary motion to the choir, nothing too close to what the choir will be doing, 'singing through the cracks', finishing on a note of tension, creating momentum and interest. It's not perfect because there are some moments of hesitation, perhaps some of the notes were not the best choice, and I know I have to go for another take. However, I do recognise some useful ideas and begin to note them on my score. In a covert way I am selecting the parts, authoring the changes and producing the alternative melody. Despite the freedom to create a part for myself, I still tread conservatively, keeping my ego in check and listening intently to the feedback from the control room.

'I like the G you finished on there,' Jared observes.

Although I later realise that Jared has incorrectly identified the last note, in this moment, I make a mental note to sing a G on the next take.

'Let's just do a bunch of them and see what comes out. We've got to remember to keep the choir in mind. I think it'll work better with the choir singing the crux of it underneath you'.

This is obviously important to Jason and so I will aim to keep this in mind during the next take.

As the instructions begin to mount, I start writing performance cues on the sheet music to keep me on track towards a selection of suitable takes. I know that the final version of this lead vocal will be 'comped', cut and pasted from the various takes, so my attention begins to move from the perspective of a larger section to that of individual phrases. Smaller modules each need to be re-interpreted and arranged to form a new vocal part.

On take 3, I break the 8-bar section down into four 2-bar phrases. The first phrase goes down, the second goes up. I'm happy with the first one. Also, bar 4 mimics the melody, which reinforces the climactic middle part of the bridge section that ends on a high A4. The third phrase gets a bit loose and gospel-like in its delivery. I'm really getting into this! When I start the fourth phrase and the first two notes don't come out right, I stop.

'Hang on, I need to work this out some more', I say.

I'm now happy with the first two phrases I want to zero in on the second half.

'I just need to work my way into that C minor 7'.

'Just on that "we've been there before" part, that's the part for me that's a bit unsure', Jared politely points out.

Jason suggests that I find another line a little higher in pitch. Picking up the score, I move over to the piano that stands in the corner of the recording room and begin annotating my plan. This is the point that I realise that the note I'm supposed to finish on is actually an F.

The suggestion that I was finishing on a G might have affected my approach to this section and caused the confusion in bars 7 and 8. There are mounting performance cues, especially in the second half of the section. But as the first half becomes more set, I'm able to start thinking about emphasising particular lyrics, voice qualities and locking in with the rhythm. It takes just a few minutes to work out a melody that I hope will capture the performance and connection that we're looking for.

On the fourth take, I do the same phrase 3 that I did earlier, but with more clarity in the note selection. I also stumble over the last three words of the section, assuming that it is just a 'slip of the tongue'. Jared says he likes it, but I'm still not happy with the third phrase. I want to sing below the melody on this phrase and I note that my A flat is not sounding well against the B flat in the melody. I change it to a G on the score and ask to 'go again', which means that I know what to do next, with no need to discuss.

As I prepare for take 5, there are changes in the governing variables, but I have made them without consultation. This is a different form of loop learning that relies on the individual reflecting on behalf of the group. It does not consider their opinion, but in the interest of studio time and in the context of relative power balance in the system, this is not an unusual happening.

I am focused mainly on hitting the G at the start of bar 6 and the F at the start of bar 8.

With no focus on my preparation, I accidentally sing the soprano line in bar 1. Bars 5 to 8 are sung accurately but carefully.

'You sang the notes right', Jared says, 'but I want you to sound like this is the hundredth time you have sung it, not the second. The struggle needs to be in conveying the meaning, not 'which word should I be singing' on which note'.

For the first time in this session I am being challenged to show expression in the performance. I have spent just ten minutes on this section, rearranging the melody and performing five takes, and now the expectation from the control room is that the notes should be set and the performance needs to come to life. Can I trust my memory to deliver the syntactic elements we have discussed as well as focus on the energy and subtle dynamics of an expressive performance?

I close my eyes and try. Accidents are mistakes that you can save. This take contained a mistake. I immediately implore, 'Let's go again'.

The disruption causes us all to laugh, and after a brief conversation on an unrelated topic, I feel elevated from the deep concentration I have been in for the past six takes to a more relaxed state. Maybe these little chats we have in between takes helps to settle me?

I close my eyes again and sing, trusting in my memory to recall the lower order syntactic performance cues and attending only to my connection with the lyrics. The seventh take is a success!

'I was really happy with that', Jared says, and continues, 'the take was perfect, but I just want another one in case I listen back and go, "wait a minute, that wasn't perfect". Do we have another option?'

I have a great feeling about this take, but I have noticed that I slightly changed the melody in bar 5 and the timing of the last note. I am happy to adopt the melodic change and say nothing about it.

I have assessed that the melodic change was an accident that adds to the performance and that the rhythmic change does not work with the choir phrasing and is therefore a mistake. I simply need to correct this and deliver another great expressive take.

As we record take 8, it's going really well, the best performance so far. Until the last phrase, where I slip up, have a mental 'lapse' and it all falls apart.

'That wasn't so perfect', chortles Jared, 'but it was great until bar 51'.

'I keep stumbling there every time', I reply. Jason points out a 'typo,' a mistake in the printed lyric at that same spot.

I hadn't even noticed the typo, but now that I am consciously aware of it, it might be easier to sing the phrase.

On the final take I get right to end with no mistakes, and Jared is happy.

'Great, man. I think we have everything we need'.

Discussion

Sociocultural aspects of production had an impact on the power balance and shifted a share of the roles to the performer. This transference of roles supports Driver's (2015) and Moorefield's (2005) arguments that roles in the studio are often not singular or clearly defined, and also exposes the level to which a vocal booth performer may be required to inhabit these roles. The singer was, at times, operating in two roles simultaneously – performer/arranger, performer/producer – which, while appearing necessary, might not be the most effective way to proceed. In situations in which the singer was attempting to accomplish more than one difficult task at one time, the number of governing variables increased, which effectively instilled more thoughts and increased cognitive processing to the point at which both tasks were compromised. By isolating particular roles within the performance, this negative outcome can be alleviated and enable the performer to complete their task(s) more effectively.

Similarly, even in familiar and comfortable surrounds, a disruption in the material and technological environment can add to the number of governing variables in use, and thus affect thoughts and performance. When one side of the headphones stops working or the balance of the mix changes significantly, attention shifts from governing variables focused on performance to feelings and beliefs that 'all is not right'. However, as is evident in this case study, such disruptions can have positive effects, as these events force changes in action strategies that can lead to desirable performance outcomes. These 'happy accidents', as they are sometimes called in the recording domain, can also result from idiosyncratic choices the singer makes (Williams, 2010), such as the novel syntactic phrasing choices adapted in takes 2 and 3.

It is apparent in this case study that when the singer tries to attend to two or more aspects of performance at a time (e.g., performance role, technical difficulties, varying the arrangement, performing the song), the number of governing variables increases to a point at which cognition is impaired and performance is undermined. Understanding the element(s) that are being attended to at any one time enables the pattern of learning and memorisation of a work to progress more efficiently.

In this case study, freedom was greeted enthusiastically by the singer as a challenge. But in a scenario in which the performer fills multiple roles, occupies a greater share of the power distribution and must develop musical structures and select novelty, freedoms need to be managed carefully and astutely to achieve effective results.

There are many traditional power arrangements within the recording studio domain. Sociotechnical processes such as the location of the recording session, organisation of space, the choice and use of equipment are usually determined by technicians rather than performers. Engineers and producers are empowered to make decisions on behalf of the performer because it is assumed that they have superior knowledge in regard to technical aspects and thus it is part of their job description. This can result in decisions that are based on the way the technicians work, a one-size-fits-all approach, to which vocal booth performers will usually defer.

However, in this case study, a number of these stereotypes were challenged by the preparedness of the control room participants to relinquish some of their power. The vocal booth performer, while initially positive about the extra authority, was clearly affected when these accompanying responsibilities compromised his overall performance. Regardless of the vocal booth performer's ability and experience, this case study suggests a correlation between efficiency and effectiveness in performance and clearly set sociocultural expectations, powers and responsibilities.

References

Driver, C. (2015). The collaborative mix: Heidegger, handlability and praxical knowledge in the recording studio [online]. In *Into the mix people, places, processes: Proceedings of the 2014 IASPM-ANZ conference*. Dunedin, NZ: International Association for the Study of Popular Music, IASPM Australia-New Zealand Branch Conference.

Moorefield, V. (2005). *The producer as composer*. Cambridge, MA: MIT Press.

Williams, A. (2010). Navigating proximities: The creative identity of the hired musician. *MEIEA Journal, 10*(1), 59–76.

Appendix 3
Case study 3: compromised music

As a professional session singer, for over 25 years my identity has remained relatively anonymous to the listening public. Yet my performances have directly influenced thousands of people through album recordings, television and movie soundtracks, advertising jingles and production music. Production music is a term used to describe music written and produced (i.e., recorded) specifically to be licensed to customers for use in film, television, radio and other media. Many of the largest music publishers in the world have production music departments that scour territories for songwriters and producers who can deliver albums of quality original material to serve this purpose. Songwriters like Phil Buckle have dedicated their time and energy to creating new product for these companies. Phil finds enjoyment and freedom in making music that evokes emotion, that might one day sell a car or a holiday or find its way onto a television series somewhere in the world. I have worked with him over the years on various albums, enjoying the challenge of delivering a lead vocal that suits that particular project's theme, anonymously connecting to the listener and sustaining this important professional outlet for Melbourne session singers.

Finding the right performance for production music can be challenging for both the performer and the producer. It begins with selecting the right session singer for the job, as Phil explains:

> It's interesting with singers, take them out of their environment and sometimes it works spectacularly and sometimes it's a bit hmmm – great singer but not the right song for that voice. They did nothing wrong, it just isn't kinda right or something. That happens quite a bit with production music, because we do so many songs. Whenever you write a song, I want it to sound like that [a particular reference track], as soon as it comes back and it's like, not as good, so the whole song has kinda taken one step down. That's disappointing. I don't have the budget to go 'I'm going to go and get another singer'. That is the thing with production music, it is

a compromise. I hate that when I want the song to be the best that it can possibly be.

I am standing in Phil's recording studio situated in the basement of a residential property in Melbourne's eastern suburbs. It is a single room that accommodates a well-appointed control console, a microphone for recording vocalists and an array of musical instruments, boxes and cases (figure A3.1). The comments above illustrate Phil's candidness and willingness to openly discuss aspects of both the business and process of recording studio production. Over the years, we have had many of these conversations. Sometimes they go on for half an hour in between takes, but they are always interesting as we share our experiences and reflect on the business we have both inhabited for the past few decades. In terms of sociocultural capital, we are fairly evenly matched. We have a similar circle of friends and have worked with many of the same people in the Australian music industry – but rarely at the same time. This is reflected in our conversation which flows freely and easily in both directions.

I understand his remark about the unsuitability of some singers for some briefs. I have witnessed it myself, and occasionally I have been that singer.

Figure A3.1 Studio of Phil Buckle. Multiple-purpose studio space.

Phil has selected me to deliver the vocal for today's session, and while I am confident, I also need to be vigilant for the sound that suits the task as well as the specific nuances that Phil is looking for in the performance. 'I want it to sound like Lady Antebellum kind of thing. It's very commercial pop with a very strong country flavour'. Phil plays me his reference track for this song, a tune called 'Need You Now' (Scott, Kelly, Haywood, & Kear, 2009), which features a duet between two members of the band. 'Just be Rod and it's going to work', Phil jokes. 'It's just the tight harmony thing. We've got the female part and we have to lock to that'.

Phil's brief helps me to forge an initial idea of the outcome he is hoping to achieve in this session – an adult contemporary style vocal performance with a vibrant and youthful sound. Therefore, I must ensure that my voice contains elements of style that belie my actual age of 49 years. Phil further explains,

> Right now, in production music we're always trying to sound contemporary – our clients are telling us we need 'that' kind of vibe. But the problem is that a lot of young singers don't have the vocal capacity to do what experienced session singers can do, so I get the tone, but the sessions are long and there are heaps of issues. When they sing unfamiliar stuff there are problems with tuning and learning melodies. I find with you, there is a lot of stuff you do intuitively.

I am comfortable with Phil's confidence in me, but today I feel a bit different, impaired by a minor surgery undergone two days ago to remove a melanoma from my back. It is not just the incision, stitches and stretched feeling in my left shoulder blade that I cannot ignore, it is the anxiety I have experienced since learning just two days ago that my health was at risk. The effect of this wound on my performance is yet to be determined, but I flag it with Phil, indicating that I might need a rest at some point in the session. Vocal booths are usually 'standing spaces', and this is no exception. In some situations, in which sessions stretch out over hours, seats are provided, but there is no space for that in Phil's room. However, I can sit behind him in the console area if I need to. The absence of a wall between us makes this passage easily accessible.

The 'booth' in this case comprises the whole room, but as always, there is space designated by the position of the microphone that dictates where I will perform. An AKG C414 XLII condenser microphone stands facing a corner of the room (Figure A3.2). This situates me with most of my back towards Phil, who is positioned behind the console. I ask about the microphone placement. Referring to the acoustics of the room, Phil says, 'I don't want it to be too dead, but I don't want it to be too live either'. 'Would it be

Figure A3.2 Studio of Phil Buckle. The placement of the microphone

better if we had eye contact?' I ask. 'I think about the singer as wanting to be off in their own world, not looking at me all the time. You might be scared to emote or make a funny face or . . . move your arms or whatever'. I find that perspective interesting. In live performance, singers want to communicate to an audience, but in the studio, they become detached from that aspect, and I wonder if that detachment is also preferable for the technicians. This is a modern production studio, where sonic isolation has given way to practicality and space restrictions. Phil explains:

> Sometimes I wish I could [isolate the singer]. Sometimes I want to listen to the mix [through the speakers]. I am not in love with headphones. I never mix on headphones, I never listen to anything on headphones. But if I was listening on the monitors, I would always be trying to sit it in the mix, I would always be trying to work on that. Because I am a songwriter, I am always trying to figure out 'is this song working?

Phil often refers to himself as a songwriter, even though he is obviously the engineer and producer is in this production. This is often the case when songwriters are multi-skilled or are forced to become multi-skilled due to the financial constraints on their business. 'Teams of people make great music', Phil states. 'But nowadays it's people in rooms with computers, like me, making compromised music. Very talented people, but they are one person and you don't get the different views'. I think about my role and how important my views become in this scenario. I am a second opinion, a willing listener and co-creator, but I know from experience that this position can also affect my principal role as a vocalist, which in itself requires a concentrated effort to deliver a suitable performance that will make the session a success and repay Phil's faith in me as a session singer.

The melody for this song has not been notated. Instead, a lyric sheet sits on the music stand to my right (Figure A3.2). 'Let me sing you this chorus', Phil says. Without a notated score, I rely on a pre-recorded guide vocal, marking the lyric sheet with annotations that represent relative changes in pitch and rhythm.

I make a comment about the sound of the female singer's voice. 'It is different to the girl from Lady Antebellum. She's got more of a Carrie Underwood sound. So do we stay in that Lady Antebellum kind of thing or is it that more traditional 'older' Nashville vibe? You can just tell me where you want my voice to sit timbre-wise'. 'I would like as much warmth as I can get', Phil replies. 'Tonal-wise, it'll be determined by the range, I guess – the verses are higher, but as much warmth as you can give me. And then we'll see what it sounds like coming back. The guy in that band, he has a distinctive voice'. 'But kind of understated', I reply. 'Yeah that's the word', Phil confirms. 'Warm and understated. The back story of this lyric is that two people are finally together and it's passionate, but you don't have to over-emote that passion. Let's stop talking and do some work!'

Multiple takes: the first chorus

There were 23 takes of the first chorus section of 'The Long Way Home'. The following presents an autoethnography of this performance, focusing on seven selected takes (starting with take 2) that summarise and distil the events of this performance, and reveal some of the skills and knowledges used by the vocal booth performer in this environment. Prior to recording, there was time set aside for learning the melody and lyrics, which was done without any concern about vocal sound or pressure to record. This period lasted approximately ten minutes and included eight separate run-throughs. Each of these involved listening on speakers to the backing track which contained both the female lead vocal and Phil's male lead vocal guide.

'She changed the melody a bit from what I gave her to sing', says Phil, 'which means we need find out what these notes are going to be . . . you can change anything that you want to change'. Once again, I am confronted with the added task of contributing to the vocal arrangement. I listen to the guide track, the chords and the female part, offering suggestions of what might work in the context of a duet. Phil tells me that he has never recorded a duet before, which places my level of experience of this type of performance above his. Nevertheless, I am more than happy to take his judgements and follow his direction for the selected part.

During the first four passes, I annotate my lyric sheet with performance cues that prompt the basic syntax of the melody in relation to the text – entry points, relative pitch changes or common notes, rhythmic changes etc.

At the point at which I begin reading my cues without further editing, Phil suggests we try singing the song with the 'cans' on. I put the headphones on and sing the chorus melody, moving towards and away from the microphone, adjusting the position of the headphone cups on my ears. I am trying to lock into the right sound and timbre – no vibrato, warm and understated. 'Let me check the position of the mic', says Phil. 'It's got a sweet spot'. As we check the height of the microphone, I try to replicate my physical performance stance, but it feels unnatural while I'm not actually singing.

'Let's see what your cans are like. I had a bit of a listen before. I've got it really compressed down here, so see what that feels like. Let me know if you want anything changed'. Phil is looking for my feedback on the sound in terms of the balance and compression in my headphones. His approach is attentive to the singer's preferences, but he also has a clear go-to plan in case the feedback is not forthcoming. 'Often singers say nothing and then you listen to the mix and it's like, this sounds awful'. I agree. A good mix is a very overlooked part of the process.

We do two more run-throughs and I pick up a red pen. 'I reckon I've got the pitch, I just need to get the inflections, onsets, tails and stuff'. Writing the interpretive cues in a different coloured pen is not something I have done before. It is a result of the research project, and the awareness of writing different types of cues in different colours is an interesting new strategy for me. I need to better understand the detail of the other singer's performance, so I can lock into it for this chorus. I mark the elements that I haven't already picked up on – dips (i.e., reverse of scoops) in the melody, and timing (notes that either anticipate or land on the beat). After two more passes, we are ready to record.

The pre-recording and learning sessions have informed me that I am following a template that includes syntactic changes set by a female singer. Phil has asked for a performance that is 'warm and understated'. While I am familiar with these terms, they can be interpreted quite broadly, so for now, I will record a benchmark version of the song, which will tell us both something about the direction of the next movement.

During the take, I focus on my acoustic environment, struggling with the sound balance in my headphones. My voice is too soft, and I have to remove an ear-cup to hear myself better. The performance is neither warm nor understated. Mistakes are made and it is sung too loudly and without expression.

'I've got you a bit hot', Phil says. 'I'll need to turn you down'. I explain that couldn't hear myself that well and I think I got a bit close to the microphone. 'Let me back off the mic and you turn me up in the cans. See if that balances things out'. As I demonstrate my understanding of relative input and output levels in a vocal recording session, I provide feedback to the engineer that will hopefully assist both his and my own performance.

'I know you're still not overly familiar with the song, but I'd say it sounds like you're singing backing vocals, not like you're singing with her. Normally I would wait a while before I would say that because we're still getting to know the song. But have a think about that'. I explain to Phil that I will try to get the cut offs the same as her – sing more together. 'Let's have a listen to her one more time'. Phil plays back the take and I listen closely to her notes and ends of phrases, comparing them to mine.

Duetting is like doubling; you need to listen closely to the other voice and reflexively adjust what you are doing to match the other voice syntactically and expressively. I also realise that I need to find my own sweet spot against the microphone, where I can deliver the right tone, input level and hear myself clearly through the headphones.

I stand a few inches back from my position on the previous take. I close my eyes – I know the words now and I feel I can put these words into my body, making them feel more human. The annotations on my lyric sheet provide cues that help me to remember what to sing, but sometimes it is just as valuable to be less concerned about a perfect representation of the work and aim for a good performance. It reminds me of something that happened earlier in the verse section of this song. Phil said, 'It's like you're hitting the word "perfect" so perfect that it's not natural enough. You could do a little slide before it and it'll be fine'. I decide to add some stylistic elements such as 'creaks' and extra breath at the ends of my phrases that I hope will make them sound more connected to the other voice. However, some mistakes and accidents still occur.

'That was a great take', says Phil. 'It felt really good. There might have been a different note in there, but I kind of don't care. It felt like – ok now we have two singers. Let's have a listen'. As we reflect on the take, Phil comments that I have a great memory for melodies. I explain that, for me, it's about predicting what is coming, then correcting where those predictions go wrong and writing these on my lyric sheet. Phil replies, 'you know, the more times you sing it, the more I get to like it. You stop doing that process of 'these are the words, this is the melody', something happens with the brain and it just starts kind of doing it. This is when I start going ok let's now just do it a few more times and see what happens'. I am grateful that Phil is so articulate about his perceptions. But regardless of the current research project, these are the types of conversations we would normally have in this space. With no physical or cultural wall to divide us, we share our ideas freely and easily.

Reflecting on this discourse, I aim to approach the next take similar to the previous take, but with less deliberate attendance to the structural cues. However, as I focus on interpretation, my technique is compromised and I begin losing the warmth, using less air and singing much louder. It is not until the end of the take that I get close where I wanted to be at the start. This deliberate change in focus has temporarily derailed my progress.

'I like the tone that just kicked in there at the end. I don't know what happened. But have a listen to that take. There are a couple of notes we still need to change'. Phil is always positive about the outcome, always focusing first on aspects that worked and allowing me to pick up my own mistakes by listening back to the take. 'I think you need to track her melody even more', he says.

> It's the way she throws away the endings of some of the words. It seems like we need to have two lead vocals and you need to have as much character as the other singer, not just supporting her and being in the background. By copying the character inflections, she has used, you get on an even level with her – that's interesting. It's almost like you have to assume her character to have some character!! Otherwise you end up being the BV singer. I reckon we've almost got it. Let's copy the way she tails of the ends of the phrases. She does something there and you're doing it really straight. Listen to the way she does 'tonight and fight'.

'She does the "creak", I say. 'Yeah, that's the character and you're the lead singer'.

I need to commit to the duet by not only matching her parts but also her presence and nuance, creating cohesion between our voices, while also making a statement about my own character or sound. I create more move-ment in the melody, tracking her notes, listening closely to her part and matching the onsets and ends of phrases, while also adding more elements of style to my sound – lip smacks, growls, in-breaths and creaks.

Phil looks happy. 'We're on it – I think we've cracked the code. That take had that thing where suddenly I become the listener and it becomes very believable. We just need to fix one word', he says. But I reply, 'give me a run at the whole thing. I am just starting to lock into it'.

I have made the decision to record another full take, exercising my power in this situation. The progress made so far is through my attention to closely tracking the duetting vocalist, but I now have permission to improvise timbre, sounds and to create textures, going beyond the 'normal' sound of my own voice. We record several more takes, as I try different voice qualities and various elements of style. Each take contains some ideas that work, but this new pathway is taking its toll on my body. Fatigue is setting in and I begin to stoop, feeling an ache where the stitches in my back are being stretched and pulled. In my quest to develop new timbral qualities, I have compromised my well-being.

Phil notices a tonal shift – 'it sounded really different'. I explain that I was adding more 'texture' – scoops and creaks – and through stooping, I might also have got closer to the mic. 'Well, the top end disappears when you go in close and you don't sound like you're singing with her anymore'. The feedback I get by continually listening to each take puts me more in touch with the changes in my sound. 'Also, can you turn the vibrato completely off? It usually makes the vocal sound younger for some reason'.

I aim to correct my posture and balance. Singers must be continually aware of this during a recording session in order to be able to sing consistently for 3 hours. In this case, it has affected my ability to control the more difficult sections and affected my vibrato and tone. I need to think about an anchor point other than my upper torso. Standing a little further back, I adjust my posture via pelvic floor anchoring, position myself at the microphone with a more upright stance, and focus on voice quality with a more open vocal tract and less air pressure. My performance incorporates fewer elements of style and has a more consistent 'warmth'. 'That shaves about 10 years off you!' Phil exclaims. 'Have a listen and tell me what you think'.

'This is where I would say let's move on, but you are one of the few singers I would trust to go again, because the takes just keep getting better'. I ask Phil what makes him think we've got it. 'As soon as I felt something, that's what you want the song to do for you'. I reply, 'OK let's go again. See if it gets better!'

I feel that all of the pieces are in place, and I just need to sing the duet – a warm and understated story about two people who find each other. I close my eyes and perform. My audience is inside my head and consists only of the voice that accompanies me, a companion that exists in my imagination. Each take develops new ideas until Phil finally announces 'best take ever! It sounds passionate to me. Let's listen'.

Discussion

The findings of this case study focus on specific methods and learning strategies that forge renewed and (mostly) improved outcomes with each repetition. The publishing company's objective to relate the song clearly to a contemporary pop act narrowed the scope for how the duet should sound. This common aspect of session singing means that performers must adopt slight changes to their timbre and elements of style that provide hints or threads to a particular reference track. This skill is not as much imitation of style as it is application of style parameters. The task also required that I follow a part that had already been recorded. The lead vocalist who performed first established a template for me to follow, which meant that I had, once again, to adapt to fit cohesively with her performance, as well as create my own sense of character as a lead vocalist. This combination of accommodation and distinctiveness was achieved by singing very 'tightly' to her vocal in terms of pitch, timing, onsets and cut-offs, but also establishing my own style parameters, changes in timbre and other participatory discrepancies (Keil, 1987) (breaths, lip smacks, growls, etc.), which characterised my performance and brought it more to the forefront of the musical texture.

Phil communicated his syntactic ideas either by singing them or by playing them on the guitar, directly intertwining language and sound (Feld, Fox, Porcello, & Samuels, 2004). With the exception of the text, instruction regarding all syntactic and expressive elements were received via aural transmission, requiring me to memorise pitch, timing and individual nuances of the melody. Such scenarios are common in session singing, giving value and power to performers who can learn a song quickly and have processes in place such as external annotations and performance cues that facilitate memory. Phil trusts that I will bring to the session my capacity to sing 'unfamiliar stuff', which sets me apart from other singers – learning melodies quickly and (usually) singing in tune.

He also knows that I can regulate my emotional responses in the studio and that he can speak freely without being concerned about 'killing the vibe' (Watson & Ward, 2013). Our shared level of curiosity about the recording process facilitates regular open discussion as well as conversation about other topics, creating a feeling of comfort and familiarity in the studio. Mutual trust and respect encourage experimentation and development of strategies, and reduce defensiveness and competition for ideas, an effect which is vital for delivering a successful outcome (Argyris, Putnam, & Smith, 1985).

In this case study, the producer felt encumbered with the responsibility of selecting the 'right' singer for the task. The session singer was subsequently required to 'be' the right singer, and thus lessen the sense of compromise that any of the participants might feel. To do this, the vocal booth performer must first understand precisely what the producer wants, and then 'cloak' themselves in the guise of the idealised performer. Session singers can 'be' themselves but must also 'be' what the recording session requires. It is a task of refinement that demands the putting away of self, while also continually asking for new innovation and subtle variations on each successive take. Thus, the vocal booth performer must bring their own self to the task, exercising their right to make decisions, but always maintaining focus on being the 'right' singer for the job.

Repetition is central to discovering these two aspects. It serves to learn the material to be sung and also, inhabit the character of the singer the producer wants you to be. During the session, these tasks were clearly separated, with the singer given freedom in the learning phase to concentrate solely on pitch/time structures, a tactic that helped greatly to emotionally regulate the recording session. This is a sociotechnical application that might appear obvious, but one that is often overlooked.

What was the 'something extra' in this case study? Recording a duet separately from the other singer requires different skills from doing it together and in-person. There is an intangible connection that you have to achieve, without which, the performance can sound disconnected and 'flat'. The act of repetition in this case study enabled me to not only learn how the song should sound, but also 'get to know' the person I was singing with. I was able to get familiar with her sound, phrasing, diction and purposefully adapt my performance to make that important connection.

References

Argyris, C., Putnam, R., & Smith, D. M. (1985). *Action science: Concepts, methods, and skills for research and intervention*. San Francisco, CA: Jossey-Bass.

Feld, S., Fox, A. A., Porcello, T., & Samuels, D. (2004). Vocal anthropology: From the music of language to the language of song. In A. Duranti (Ed.), *A companion to linguistic anthropology* (pp. 321–345). Malden, MA; Oxford: Blackwell.

Keil, C. (1987). Participatory discrepancies and the power of music. *Cultural Anthropology*, *2*(3), 275–283.

Scott, H., Kelley, C., Haywood, D., & Kear, J. (2009). Need You Now. On *Need You Now* [CD]. Capitol Nashville: Parlophone.

Watson, A., & Ward, J. (2013). Creating the right "vibe": Emotional labour and musical performance in the recording studio. *Environment and Planning A*, *45*(12), 2904–2918. Retrieved from https://dspace.lboro.ac.uk/dspace-jspui/bit stream/2134/21404/1/EPA45-208Emotionallabourintherecordingstudio-FINAL. pdf

Index

Note: Page numbers in *italics* indicate a figure and page numbers in **bold** indicate a table on the corresponding page. Page numbers followed by "n" indicate a note.

Milton Keynes UK
Ingram Content Group UK Ltd.
UKHW022358061024
449327UK00031B/2561